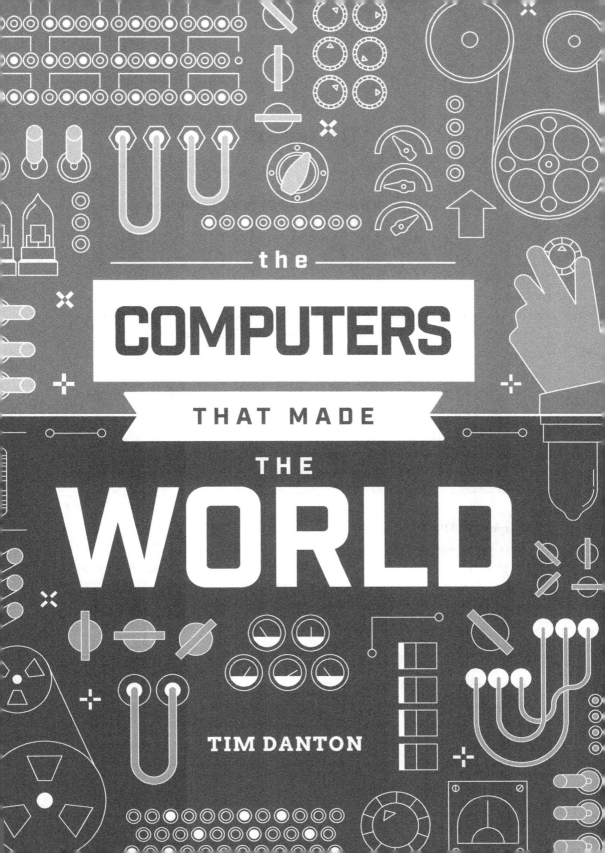

Colophon

Raspberry Pi is an affordable way to do something useful, or to do something fun

Democratising technology – providing access to tools – has been our motivation since the Raspberry Pi project began. By driving down the cost of general-purpose computing to below $5, we've opened up the ability for anybody to use computers in projects that used to require prohibitive amounts of capital. Today, with barriers to entry being removed, we see Raspberry Pi computers being used everywhere, from interactive museum exhibits and schools to national postal sorting offices and government call centres. Kitchen table businesses all over the world have been able to scale and find success in a way that just wasn't possible in a world where integrating technology meant spending large sums on laptops and PCs.

Raspberry Pi removes the high entry cost to computing for people across all demographics: while children can benefit from a computing education that previously wasn't open to them, so too can the many adults who have historically been priced out of using computers for enterprise, entertainment, and creativity. Raspberry Pi eliminates those barriers.

Raspberry Pi Press
store.rpipress.cc

Raspberry Pi Press is your essential bookshelf for all things computing, gaming, and hands-on making. As the official publishing imprint of Raspberry Pi Ltd, we publish a range of titles to help you make the most of your Raspberry Pi hardware. Whether you're building a PC or building a cabinet, you can discover your passion, learn new skills, and make awesome stuff with our extensive collection of books and magazines.

Raspberry Pi Official Magazine
magazine.raspberrypi.com

Raspberry Pi Official Magazine is a monthly publication for makers, engineers, and enthusiasts who love to create with electronics and computer technology. Each issue is packed with Raspberry Pi–themed projects, tutorials, how-to guides, and the latest community news and events.

Published by Raspberry Pi
Trading Ltd, 194 Cambridge Science
Park, Milton Road, Cambridge,
Cambridgeshire, CB4 0AB.

Raspberry Pi Ireland Ltd, 3 Dublin Landings, D01 C4E0, compliance@raspberrypi.com

Publishing Director
Brian Jepson

Sub Editor
Nicola King

Illustrations
Sam Alder with Brian O Halloran

Technical Editing
Mike Thompson

Editor
Phil King

Design
Critical Media

CEO
Eben Upton

Special thanks to: Penny Ahlstrand at the Computer History Museum; Ed Eckert at Nokia Bell Labs

ISBN 978-1-916868-07-6

July 2025: First Edition

The publisher, and contributors accept no responsibility in respect of any omissions or errors relating to goods, products or services referred to or advertised in this book. Except where otherwise noted, the content of this book is licensed under a Creative Commons Attribution-NonCommercial-ShareAlike 3.0 Unported (CC BY-NC-SA 3.0).

Contents

Timeline – From 1936 to 1950 viii

Introduction – Charles Babbage – inventor of the first mechanical computer x

Chapter 1: ABC – The Atanasoff-Berry computer 2

Chapter 2: Zuse Z3 – An early electromechanical computer. 20

Chapter 3: Complex Number Calculator – Building the foundations of digital computers . 38

Chapter 4: Colossus – Code-breaking computer that helped win a World War. 58

Chapter 5: Harvard Mark I – Another pioneering electromechanical computer 86

Chapter 6: ENIAC – The first programmable digital computer. 108

Chapter 7: Manchester Baby – The first electronic stored-program computer 128

Chapter 8: EDSAC – Pioneering British computer 150

Chapter 9: EDVAC, UNIVAC, & Princeton IAS – Three computers with shared origins. . . 166

Chapter 10: Pilot ACE – Vacuum-tube powered early computing 202

Chapter 11: What happened next – The growth of commercial computing 230

Timeline – From 1951 to 1965 246

Index. 248

Acknowledgements

This book would not have been possible without the help of many people who were kind enough to help me check facts, provide guidance and submit themselves to interrogation.

In order of the chapters, my heartfelt gratitude to both John Gustafson and Charles Shorb. Speaking to you both made the story of the ABC come alive, and I hope your contributions will do the same for anyone who reads the story behind that remarkable computer. It was also a real privilege to interview computer scientist and historian Raúl Rojas as part of my research on the Zuse Z3 computer (and its siblings).

For many of the chapters, such as George Stibitz's Complex Number Calculator, the ENIAC and the EDVAC, I benefited from the wisdom of interviewers who spoke to the people involved in these computers' creation and captured their stories in oral histories. You will find many references to their work here, whether that's Nancy Stern, William Aspray or Thomas Lean.

If the Colossus story does any justice to Tommy Flowers, Bill Tutte, and all the other amazing engineers and code-breakers that brought this machine to life, that's thanks to Jack Copeland. Jack is a historian, professor and an incredibly patient man whose eyes are sharper than a hawk's when it comes to clarifying meaning and spotting mistakes. Any mistakes remaining rest on my shoulders.

Jack again deserves much of the credit for the chapter on the Pilot ACE. This time not only for his direct help, which was considerable, but because much of my research was based on the site he helps to maintain at **alanturing.net**. This is an amazing resource for anyone who would like to know more about Turing – or you can buy one of Jack's many books, which include *The Essential Turing*.

It was also a great pleasure to interview Simon Lavington as part of my research on the Manchester Baby and the computers it grew up to be. Not only for that, but for spotting several mistakes that might otherwise have slipped through. Simon is the author of the several books that cover early British computers, and I can only hope that some of that expertise has made its way onto the page. My further gratitude goes to Jim Miles, Emeritus Professor of Computer Engineering at the University of Manchester, for his time and feedback on making that chapter the best it could be.

As with any book, there are many people behind the scenes who turn words in a document into something worth reading. Lee Allen at Critical Media has done an exceptional job with the design, and I can't thank Phil King enough for sourcing the vast majority of the images alongside editing my wayward prose. My thanks also to Brian Jepson at the Raspberry Pi Press for his incredible patience – let's not talk about missed deadlines – and to Russell Barnes for commissioning this book way back in 2022.

There are two institutions that also deserve mention. One is the National Museum of Computing (located on the same site as Bletchley Park), which is where you'll find the replica of a Colossus. With any luck, you'll bump into one of the many volunteers who shared their time and expertise with me during my visit. Particular thanks to Charles Coultas and the ever-effervescent museum director, Jacqui Garrad – who was also a huge help when writing my previous book, *The Computers That Made Britain* (ISBN 978-1912047857).

Finally, this book wouldn't have been possible without the support of my family – Kim, Fraser, Rowan, and Penelope – and the huge building that stole me away from them for weeks at a time. The British Library is an incredible institution that is worth visiting even if you don't intend to read its unmatched collection of books. Through the library, and its brilliant staff, I read the original 'Colossus' report by Brian Randell, newspapers on microfiche from the 1940s, fragile records of early mechanical computers, and over 100 other books that have contributed to the knowledge packed within this modest tome.

And a final thanks to you, for reading these words. I hope you enjoy the rest of the book.

Tim Danton

The Computers That Made the World
TIMELINE

1936
Alan Turing publishes 'On Computable Numbers', introduces concept of Universal Turing Machine

1937
George Stibitz builds Model K relay-based adder in his kitchen

1938
In Germany, Konrad Zuse completes the Z1 mechanical computer, a forerunner to the Z3

1939
3 September: Britain and France declare war on Germany, marking official start of World War II

Complex Number Calculator
Bell Labs' electromagnetic-relay-based computer (later known as the Model I) is completed in October

1940
Complex Number Computer demonstrated to world, with John von Neumann and John Mauchly in attendance

1941
John Mauchly spends several days with Dr Atanasoff and what would become the ABC

7 December: USA declares war on Japan following attack on Pearl Harbor

Z3
Zuse completes the Z3 computer, using electromagnetic relays for storage

1942
First bombe, used by British to decrypt Nazi Enigma messages, comes into action

ABC
Dr John Atanasoff and Clifford Berry complete work on special-purpose computer at Iowa State University

EARLY DAYS

1944

▶ **Colossus**
First Colossus (nine more would follow), a special-purpose valve-based computer, arrives at Bletchley Park

▶ **Harvard Mark 1**
Howard Aiken's electromechanical Harvard Mark 1 is completed in partnership with IBM

1945

John von Neumann publishes first draft on EDVAC. Alan Turing writes proposal for ACE

End of the war: VE Day (Victory in Europe Day) 8 May, VJ Day on 15 August ends hostilities in Pacific

1946

Moore Lectures take place at University of Pennsylvania to give attendees 'blueprint' on how to build a computer

▶ **ANIAC**
First large-scale digital computer unveiled to world and amazes due to its sheer speed of calculations

1948

▶ **Manchester Baby**
First program to be stored within computer is run. Soon expanded to become Manchester Mark I

1949

▶ **EDSAC**
Cambridge-based computer, inspired by von Neumann and Moore School lectures, ticks into action

▶ **BINAC**
Eckert and Mauchly, creators of the ENIAC, deliver Binary Automatic Computer to Northrop Aircraft

▶ **EDVAC**
Delivered to the Ballistics Laboratory in 1949, but wouldn't run for prolonged periods until 1951

1950

▶ **Pilot ACE**
Alan Turing's plans for a universal computer finally arrive – in relatively 'small' form – in the Pilot ACE

POST-WAR COMPUTERS

Timeline   ix

Charles Babbage's Analytical Engine

"It is unworthy of excellent men to
lose hours like slaves in the labour of
calculation which could safely be relegated
to anyone else if machines were used."
Leibniz, 1671

Groucho Marx was very nearly right when he said there are two certainties in life. There are indeed two, but they aren't death and taxes: they're that humans have an innate need to count and that we all make mistakes.

The trouble is that mistakes cost lives. Anyone working in the fields of engineering or navigation in the 1800s relied on printed mathematical tables that contained pre-calculated answers to equations. This meant they didn't need to work out the results by hand, and theoretically the answers would be far more accurate than someone attempting to work them out on demand.

Charles Babbage, 1860
Image: Public Domain

Theoretically: in reality, the tables contained errors almost impossible to spot that could cost lives and livelihoods.

Not that Charles Babbage needed to worry about putting his own life at risk or poverty. With a generous inheritance and formidable intellect, he was an enfant terrible of the academic world, scorned by some, admired by others. By the time 1821 rolled around, when he had just turned 30, he had many achievements to his name: co-founder of the Analytical Society to promote continental advances in mathematics in England, Fellow of the Royal Society for his contributions to science and maths, and a founding member of the Royal Astronomical Society.

He and his friend John Herschel had taken on the job of preparing a set of star tables for the new society the previous year. "My friend Herschel, calling upon me, brought with him the calculations of the (human) computers, and we commenced the tedious process of verification," Babbage is quoted as saying in one account.[1] "After a time many discrepancies occurred, and at one point these discordances were

---

[1] Charles Babbage quoted in *Memoir of the Life and Labours of the Late Charles Babbage, Esq* by HW Buxton (reprinted by The MIT Press, 1988, ISBN 978-0262022699)

so numerous that I exclaimed 'I wish to God these calculations had been executed by steam'."

From this exclamation – so the story goes – came the idea of creating a machine that could do exactly that. Specifically, to create complex logarithm 'star tables' to aid navigation.

He called the machine the Difference Engine because, rather than attempt to perform a complex equation for each value, the machine would build upon the previous result and add the difference. It was an idea that Babbage borrowed from a visit to Paris in 1819, where they were remaking their tables on an unprecedented scale after Napoleon decreed that France should move from imperial measurements (pounds and ounces, inches and feet) to metric.

By creating a machine to perform this work, Babbage would also bypass the errors inevitably made when transcribing results. Now all he had to do was design and build it. The design was the far simpler affair of the two, with Babbage designing a complex web of pinions, gears, and cogged wheels that would complete the task. His paper on the subject, 'On the Theoretical Principles of the Machinery for Calculating Tables', won the Royal Society's first ever Gold Medal – its highest accolade.

He had the design, he had the backing of the Royal Society. Now all Babbage needed was the money. Here, he turned to the British government, who awarded him £1500 (which conveniently translates into roughly £150,000 today) towards the project. In tandem with so many stories of computers in the rest of this book, that rough costing turned out to be woefully low.

The Difference Engine was a precision instrument even by today's standards, and Babbage had high ambitions: to calculate results up to 20 digits rather than the then standard six digits, and there needed to be somewhere to store calculations. A basic mechanical calculator had been invented before, but this was several orders of magnitude more complicated. Take its size alone: the Difference Engine would be big enough to fill a room.

Despite the work of one of Britain's most talented machinists, Joseph Clement, the challenge proved too much. After ten years of effort and with nothing to show for their investment other than a collection of completed parts and pleas for more money, the government said enough: no more funds. Clement also walked away from the

The Difference Engine in the Science Museum, constructed based on the plans for Babbage's Difference Engine No. 2
Image: Geni, CC BY-SA 4.0

project, taking Babbage's plans with him as hostage until Babbage gave him money he was still owed.

This enforced break from the project may have been a piece of great fortune. It gave Babbage time and space away from the Difference Engine, and when he was eventually reunited with his plans – having finally agreed a settlement with Clement after 16 months[2] – and started looking through them, he came up with an even better idea.

The idea would become known as the Analytical Engine, and it was a piece of thinking a century before its time. This wouldn't be a single-purpose machine like the Difference Engine. You could change its function – reprogram it, in modern parlance – to perform any task you liked.

To describe the Analytical Engine's main elements is to echo what makes up a modern-day computer. Babbage conceived of four key components: a mill, store,

---

[2] Doron Swade, *The Cogwheel Brain* (Little, Brown & Company, 2000, ISBN 978-0316648479), p91

Introduction   xiii

reader, and printer. The mill was where calculations took place, like today's CPU or the accumulators in early digital computers. The store is where information would be held before it was processed. And the reader and printer hardly need any introduction, being equivalent to the input and output devices that we are all accustomed to.

Again, Babbage didn't hold back when it came to ambition: his store would hold a thousand numbers up to 50 digits long. To put that into perspective, even Alan Turing wasn't that optimistic when setting out his plans for the ACE. Babbage would also create an automatic printer to avoid any possibility of human mistakes. For a reader, he would use punched cards, building upon technology already created by Joseph-Marie Jacquard for the loom.[3] But rather than use these to weave multicoloured patterns on fabric, they would hold instructions and data.

Here, we shall introduce Ada Lovelace (née Augusta Ada Byron) for the first time. Famously, she was the daughter of Lord Byron,[4] poet and lothario. Ada's mother, Annabella Milbanke Byron, legally separated from her husband when Ada was only two months old over concerns about Byron's mental health, his relationship with his half-sister, and rumours over his sexuality; Byron left the country, never to return and never to see his daughter again.

Annabella was a leading advocate for education, eventually establishing a school in London for the underprivileged, and she gave her intelligent daughter every chance to thrive through private tutors. Tutors who she largely outgrew, eventually resorting to teaching herself through books.

Ada married Baron William King in July 1835 at the age of 19, becoming Countess of Lovelace three years later when he was created an Earl. By this point, she had already met Babbage and become intrigued by his machines. The two became firm friends, and it was Babbage she turned to when seeking guidance for a tutor who could help guide her. He suggested Augustus De Morgan, a mathematician and logician best known for De Morgan's laws, who would also have given her insights into the workings of Babbage's Analytical Engine.

---

[3] In 1804, Joseph-Marie Jacquard transformed the weaving industry when he patented a method of controlling a loom using cards that were punched through with holes. The hole patterns dictated the weaving pattern, and could be swapped with other cards depending on the weave's design.

[4] His full name was George Gordon Byron, 6th Baron Byron. In the British peerage system, the ranking (from top to bottom) goes Duke, Marquess, Earl, Viscount, and Baron. Titles are either inherited or given by the ruling monarch. 'Lord' is a generic word that can be applied to any peerage; so, rather than be called Baron Byron, the poet preferred Lord Byron. We can understand why.

Despite being a prolific writer, Babbage never set down his own description of the Analytical Engine on paper. The first printed account came in 1842, when Italian mathematician Luigi Menabrea published 'Notions sur la machine analytique de M. Charles Babbage' in a Swiss journal.[5] Menabrea had attended a presentation by Babbage in Turin, introducing the assembled scientists to radical concepts such as conditional branching, and it took Menabrea almost two years to complete his 23-page article.[6]

Ada took on the task of not only translating this article into English but adding her own copious notes. The end result was published in London a year later under the title 'Sketch of the Analytical Engine invented by Charles Babbage Esq.' with an additional 41 pages simply titled 'Notes by the translator'.[7] The only hint of the translator's identity being an 'A.L.L.' at the end; a typesetter's error, as those are not Augusta Ada Lovelace's initials.

There is one particular section that brings the Analytical Engine to life: "We may say most aptly that the Analytical Engine weaves algebraic patterns just as the Jacquard-loom weaves flowers and leaves," wrote Ada. There is a hint of poetry there, as we might expect from Lord Byron's daughter, but also the crucial point that Babbage's invention was far more than an accumulator of numbers.

The translation was collaborative, with much correspondence between Babbage and Lovelace during its creation. Ada's crowning glory, which earns her the title of 'world's first programmer' in many people's eyes, is her sequence of mathematical operations that could be performed on the Analytical Engine to calculate Bernoulli numbers. Even today, it resembles code.

She also looked beyond mathematics: "Suppose, for instance, that the fundamental relations of pitched sounds in the science of harmony and of musical composition were susceptible of such expression and adaptations," Ada wrote, "the engine might compose elaborate and scientific pieces of music of any degree of complexity or

---

[5] Luigi Federico Menabrea, 'Notions sur la machine analytique de M. Charles Babbage', *Bibliothèque universelle de Genève*, No. 82, October 1842, pp352-76
[6] See 'Luigi Menabrea Publishes the First Computer Programs, Designed for Babbage's Analytical Engine. Ada Lovelace Translates them Into English', on Jerry Norman's **HistoryofInformation.com**, for more background on Menabrea and this event: **historyofinformation.com/detail.php?entryid=546**
[7] *Scientific Memoirs Selected from the Transactions of Foreign Academies of Science and Learned Societies and from Foreign Journals*, edited by Richard Taylor, volume III, 1843, pp691-731

extent." Tragically, this was to be her last big contribution: she died at the age of 36, probably from cancer; she was buried next to Lord Byron, at her request.

Despite all the positive coverage for the Analytical Engine, it was destined to never be built. The machine was even more complicated than the Difference Engine and the British government simply wasn't interested in spending any more money on Babbage's hare-brained schemes.

Could it have been built using technology from that era? In his 22-page article that explores the Analytical Engine in great detail, Allan Bromley concluded that the answer was yes. "Analyses such as these [on machining accuracy and weights] lead me to believe that the Analytical Engine could have been built with the technology at Babbage's disposal, although the work would undoubtedly have been demanding and expensive," he wrote.[8]

What's certainly true is that things were much easier for Howard Aiken, creator of the Harvard Mark I, who once joked that, "if Babbage had lived 75 years later I would have been out of a job." It's hard to argue with that conclusion (although Babbage may have got distracted by a new invention instead). After all, electricity is a much friendlier supplier of power than steam, and despite Bromley's optimism it's worth noting that even the failed attempt at creating the Difference Engine pushed the boundaries of machining at the time.

Babbage eventually gave up on the idea of the Analytical Engine, deciding instead to design a simplified version of the Difference Engine. Again, he never made this machine, but a real-size replica of the Difference Engine No. 2 based on Babbage's designs can be found on display at the Science Museum, London. You'll also find another Difference Engine there, but this one was created in Babbage's lifetime and based on his first design. It was made by a Swedish inventor and his son: Georg and Edvard Scheutz. Babbage even saw his creation when they brought it to London, where it went on display at the Royal Society. It was used to generate and print mathematical tables, but sadly proved temperamental.

The tragic side effect on the Scheutz family and Babbage was that building mathematical engines brought neither happiness nor wealth. The Swedish pair would die bankrupt, while Babbage grew gradually more bitter about the government's

---

[8] Allan G Bromley, 'Charles Babbage's Analytical Engine, 1838', in *IEEE Annals of the History of Computing*, Vol 4, No 3, July 1982, p204

failure to back his machines and his failure to bring his inventions to life. Although – and those of a squeamish nature should look away now – Michael Williams, editor-in-chief of the *IEEE Annals of the History of Computing*, gives context to Babbage's reported irascible nature in later life by the string of medical conditions he was suffering from: "Who would not be 'crusty' with kidneys and urinary tracts and arteries such as these?" Williams wrote.

Nor should we simply write off Charles Babbage as simply a man before his time. He achieved incredible things during his life, with accomplishments covering everything from the invention of the ophthalmoscope to a 'black box' to help detect the reason for train accidents to proposing a scientific method for using ring dating to determine the age of trees.

There is one other factor to consider, one that is driven home by the stories of the computers elsewhere in this book. Ultimately, you can argue, it wasn't the lack of technology that halted the creation of the Analytical Engine. You also need a driving force. Money, almost without limit. These conditions are rarely found in peacetime: it took war to drive the development of the ENIAC, and Colossus, and without this how long might we have had to wait for truly electronic computers?

1997 replica of the
Atanasoff–Berry Computer
at the Durham Center,
Iowa State University

Image: Nyanchew, CC BY 2.0

ABC
(Atanasoff-Berry Computer)

As difficult as ABC: designing the first electronic digital computer

John Atanasoff reveals much of his character with one simple anecdote in his autobiographical essay, 'The Beginning'.[1] It all started in 1913, he explains, when his father bought a new slide rule that he ended up giving to his son. "In two weeks or thereabouts, I could solve most problems with it. Could you imagine how a boy of nine-plus, with baseball on his mind, could be transformed by this knowledge?"

We think it's safe to say that an ordinary child of nine would not be transformed. But Atanasoff was no ordinary child, and he would grow up to be quite an extraordinary adult.

The journey from slide rule to the world's first electronic computer took a shade under three decades, with stops along the way to study electrical engineering at the University of Florida and earn a master's degree in mathematics at Iowa State College (now Iowa State University).

It was whilst working on his PhD thesis, at the University of Wisconsin, that Atanasoff first grew frustrated with existing methods of calculation. "This work demanded long calculations with a table computing machine/tabulator to solve the equation of Shreudinger [Schrödinger]," he wrote. "Such calculations required many weeks of hard work on a desk calculator such as the Monroe, which was all that was available at the time. I was also impressed that the process of approximating the solution of partial differential equations required a great many calculations, a fact that ultimately motivated my work in automatic computing."

This was in 1930. After returning to Iowa State College, where he would soon be made assistant professor of mathematics and physics, he and a colleague ingeniously adapted a college-owned IBM tabulator so that it could solve a set of linear equations.[2] As Atanasoff described it, the tabulator was "the largest 'computer' of the day". These were huge machines that performed operations based on data held on punch cards; a technology with roots trailing back to Jacquard's loom in the early 1800s.

Although the tabulator slashed calculation times compared to the Monroe, John Atanasoff realised that the days of such clunky, mechanical machines were coming to an end. He even wrote to IBM to share his ideas, but the company was not interested

---

[1] 'The Beginning, John Atanasoff', in Dimitar Shishkov (ed.), *John Atanasoff: The Father of the Computer* (TANGRA TanNakRa, 2001, ISBN 978-9549942248), pp69-101

[2] Anyone curious as to how Atanasoff achieved this can read the eleven-page paper he wrote that detailed the process, which is digitised at **digitalcollections.lib.iastate.edu**. Search for 'Solution of Systems of Linear Equations by the Use of Punched Card Equipment'.

in the thoughts of an assistant professor from a provincial college. If anything, it appears to have had disdain. One internal IBM memo, which came to light many years later, stated simply and decisively: "Atanasoff should not be allowed near the tabulator".[3]

Unaware of IBM's attitude, Atanasoff continued to correspond with the company to share his ideas of how the tabulator could be improved by embracing electronic computing. He only stopped when he received a letter from IBM categorically stating that it was not interested in building electronic computers. A letter that caused him to remark at the time that he should frame it; he could see the future far more clearly than IBM's executives.

John Atanasoff
Image: Eye Steel Film, CC BY 2.0

Even if Atanasoff could have persuaded IBM to make this jump into the unknown, it seems unlikely that he would have waited. In 1985, 35 years into their marriage, his second wife Alice wrote: "Living with a man who is constantly infatuated with strange ideas is quite exciting. It is never boring. Sometimes I tell him that other people go and buy what they need but he wants to invent his own machine, to build some device."[4]

This character trait continued throughout Atanasoff's life. When he retired, he not only designed his own house for a newly acquired 200-acre farm, but also a mechanical tree planter. Using this, he planted an astonishing 36,000 trees.[5]

With hindsight, it seems obvious that Atanasoff's problem-solving nature, background in electrical engineering, and frustrating experiences with mechanical computing devices would ultimately lead to his invention of an electrical computer. But this oversimplifies things: one aspect of Atanasoff's genius was his ability to not only recognise the tabulators' inherent limitations, but to analyse the problem and

---

[3] Nikolay Bonchev, 'The Route to the Invention', in Dimitar Shishkov (ed.), *John Atanasoff, The Father of the Computer*, p44
[4] Alice Atanasoff, 'Living with a man who is constantly infatuated with strange ideas is quite exciting', reprinted in Dimitar Shishkov (ed.), *John Atanasoff: The Father of the Computer*, p171
[5] Tammara Burton, *World Changer: Atanasoff and the Computer* (TANGRA TanNakRa, 2006, ISBN 978-9549942941), p173

then break it down into component parts. If anything, his biggest problem – ironically mirroring the equation sets that he was trying to solve – was the number of unknowns. Put simply, he had too many options to choose from.

"I thought I knew how a computer should work," he wrote in the mid-1980s.[6] "First it would have to add and subtract, and later one could compound these operations into multiplication and division... From the start, I was interested in carry-over;[7] it is the crux of the digital method."

In contrast to the ENIAC and Harvard Mark I, completed almost a decade later, Atanasoff embraced the use of binary rather than decimal numbers. "A digital computer also requires some entity to represent numbers," he wrote in the same article. "While historically this system used numbers to the base 10, we intend no such restriction, for in theory any greater than unity can be used as the base. My own device... and most modern computers use the base 2."

He added: "In looking over the 1936 art in computing, I had become convinced that a new computer should provide for a much larger retention of data. Almost at the start, I called this 'memory'. The word seemed natural to me, as I suppose it did to others since it is still in use in a wide field including computers."

Atanasoff had solved, in theory, a couple of fundamental problems that stood in the way of automatic computers. What he didn't have was a blueprint for a working computer itself. Now we fast-forward to a courtroom in 1971. By this time, computers were big business, and the Sperry Rand company, the owner of electronic computing patents based on the ENIAC, was demanding huge royalties. To be precise, 1.5% of the cost of any new computer, payable by every manufacturer.

Every manufacturer, that is, except IBM. In 1956, it had struck a secret deal with Sperry Rand to share their intellectual property across the newly emerging field of electronic data processing. Sperry Rand had the lion's share of the patents, including the so-called ENIAC patent, which is why IBM paid $10 million as part of the deal.

Independent observers, had there been any, may well have commented that this was a small amount to pay for such a pivotal piece of IP. Why did Sperry Rand accept such a low figure? The answer almost certainly stems from a visit to Atanasoff

---

[6] John Atanasoff, 'Advent of Electronic Digital Computing', in *IEEE Annals of the History of Computing*, Vol 6, Issue 3, July–Sept 1984, p237

[7] Atanasoff is referring to the challenge of 'carrying' values across multiple digits during addition and subtraction, a fundamental aspect of performing complex calculations.

A monument to John Atanasoff in Sofia, Bulgaria
Image: Nickolay Angelow, CC BY-SA 2.5

by an IBM patent attorney, AJ Etienne, in June 1954. According to Atanasoff's granddaughter, Etienne said: "If you will help us, we will break the Mauchly-Eckert computer patent; it was derived from you." But that was the last Atanasoff heard from the lawyer.

It was only years later, when the matter came to trial, that Atanasoff realised the discussions went no further because IBM had secretly settled with Sperry Rand. The obvious but unproven interpretation being that IBM had used its knowledge of 'prior art', particularly relating to the ABC's memory system, as a negotiation tool. Rather than test the argument in court, how much better for both parties to settle the dispute behind closed doors?

When Honeywell bought General Electric's computer division in 1970, it had no knowledge of Atanasoff's computer or the secret cross-licensing deal between two of its arch-rivals. What it did know is that Sperry Rand wanted $250 million in royalties, equivalent to $2.1 billion in 2025. A figure it later reduced to $20 million, compared to the $150 million it was asking from six other companies.[8]

Honeywell decided the time was ripe to challenge this patent. Part of its argument was that John Mauchly – one of the co-creators of the ENIAC – had based a significant amount of that computer's design on the ABC. As Atanasoff's son put it in a 2010 article:[9] "John Mauchly visited Iowa State in 1940–41 and illegally borrowed some of the ideas in the ENIAC development and patent."

Honeywell enlisted Atanasoff as its key witness in the trial. "Well, I remember that the winter of 1937 was a desperate one for me," he recalled under oath in June 1971,

---

[8] Alice R Burks and Arthur W Burks, *The First Electronic Computer: The Atanasoff Story* (The University of Michigan Press, 1988, ISBN 978-0472100903), p197

[9] 'John Atanasoff and Clifford Berry: Inventing the ABC, A Benchmark Digital Computer', in *Electronic Design* magazine, 12.09.10, p59

"because I had this problem and I had outlined my objectives [for the computer] but nothing was happening, and as the winter deepened my despair grew… we come to a day in the middle of winter when I went to the office intending to spend the evening trying to resolve some of these questions and I was in such a mental state that no resolution was possible.

"I went out to my automobile, got in and started driving over the good highways of Iowa at a high rate of speed." At that point, Iowa was a dry state, and it's unlikely to be a coincidence that he was heading into Illinois for a drop of liquid inspiration. The long drive also allowed his mind to mull over complicated problems, a method he had used before. After "several hours" he arrived at a bar in Rock Island, which he describes more colourfully as a "speakeasy" elsewhere. Some of the nature of the place is reflected in the fact that it featured a secret tunnel for escaping cop raids.

After sitting down and ordering a drink, Atanasoff said that "my thoughts turned again to computing machines… During this evening in the tavern, I generated within my mind the possibility of regenerative memory. I called it 'jogging' at the time."

Three problems down, one to go. So, with much poetic licence, let us imagine Atanasoff silently shouting "Eureka!" as the final element fell into place. "During the same evening, I gained an initial concept of what is today called the 'logic circuits'. That is, a non-ratcheting approach[10] to the interaction between two memory units, or, as I called them in those days, 'abaci'. I visualised a black box which would have the following action: suppose the state of abacus one and the state of abacus two would pass into the box; then the black box would yield the correct results."

He added: "During that evening in the Illinois roadhouse, I made four decisions for my computer project:
- I would use electricity and electronics as the media for the computer.
- In spite of custom, I would use base-2 numbers (binary) for my computer.
- I would use condensers for memory, but 'regenerate' to avoid lapse.
- I would compute by direct logical action, not by enumeration."

After a year of experimenting with jogging and logic circuits (thank goodness for those years studying electrical engineering), Atanasoff was confident enough in his

---

[10] Unlike a ratcheting device, where motion can only happen in one direction, here Atanasoff envisaged a device where movement could happen in either direction.

idea to build it. Now all he needed was money and an assistant. A grant from the dean of the graduate school provided the first, an accidental meeting with a professor of electrical engineering his second: "I have your man," the professor replied, after he and Atanasoff bumped into one another on campus. "Clifford Berry."

A lazy biographer might argue that Berry was Watson to Atanasoff's Holmes, but that would ignore the younger man's intellectual prowess. "Atanasoff's propensity for coming up with a succession of new ideas while working on projects had overwhelmed a number of his graduate students," wrote Tammara Burton, his granddaughter, in her biography *World Changer: Atanasoff and the Computer*,[11] "but Berry had no problem keeping pace with Atanasoff. This ability earned Atanasoff's unqualified respect and admiration and provided the basis for an ideal working relationship and a lifelong friendship."

Atanasoff always heaped praise upon Berry. To the extent that when he realised that the computer needed a name for the sake of posterity, he decided upon the Atanasoff-Berry Computer. Although the beauty of the ABC initials may, just perhaps, have been another reason.

The team formed, the money in place, Atanasoff and Berry set to work on the prototype in early autumn 1939. Atanasoff had a theoretical blueprint in mind, but putting theory into practice meant building and testing every component.

One of the most crucial was the ABC's memory. "I chose condensers (or capacitors) as the element for memory, because a condenser can give a good voltage to actuate a vacuum tube," he wrote, "and because the vacuum tube will give enough voltage to charge the condenser."

This was the concept of 'jogging' in electrical form, and a concept that we still use to this day in computers in the form of memory chips (dynamic random access memory, or DRAM). Just like the ABC's condensers, this needs a constant top-up supply of electricity to stay alive. Or, to keep jogging. "Jogging is reminiscent of the little boy going to the grocery store and reciting, 'a dozen eggs, a pound of butter etc.'," to quote Atanasoff once more. "Over and over, hoping to arrive at the store before his memory has failed."

Atanasoff remembers the prototype being finished in November 1939, but according to materials submitted in the court case, the first demonstration happened a month earlier. Either way, Atanasoff's 'crude device' proved a success.

---

[11] Tammara Burton, *World Changer: Atanasoff and the Computer*, p100

The prototype counts so many firsts that it seems almost unbelievable that it was created by two men, neither experts in this field, in a matter of months. It used vacuum tubes (valves) rather than electro-mechanical switches. It used binary rather than decimal. It used logic systems. It used capacitors to create memory. It took the idea of jogging from theory into practice. It was, quite simply, brilliant.

But the prototype was also extremely limited. It could only add or subtract numbers, and not even big ones. The wheel mechanism contained two rings of 25 condensers each, meaning that it could represent numbers up to 25 base-2 places, equivalent to about eight decimal places.

Atanasoff and Berry knew they had made a breakthrough by creating this prototype, but to turn the proof-of-concept into a computer would take money. More money than they could raise directly from Iowa State College, despite it generously awarding a further $810 grant on completion of the prototype.

This was enough for Atanasoff and Berry to build the frame of the computer. It was roughly the size of an office desk, six feet across, and at roughly three feet wide it was a couple of inches wider than a typical door frame; a decision that would cause a fatal logistical problem further down the line. The two men worked fervently on the project whenever they could in the early months of 1940.

By this point, Atanasoff realised this was a significant invention that required patenting. So, over the course of weeks, he authored a paper called 'Computing Machine for the Solution of Large Systems of Linear Algebraic Equations'.[12] It was both the basis of a patent submission – a copy would eventually be sent to a patent attorney – and a begging letter for $5330. This, he estimated, would cover the cost of construction, materials, revisions, and two research assistants to help.

It is always hard to convert such sums into something meaningful today, but $5330 inflates to around $120,000 in 2025 terms. Bearing in mind the ambition of what Atanasoff wanted to achieve, that's a modest sum. As he signed off the report himself: "…it would have seemed absolutely impossible to the writer two years ago to have designed and constructed a computing machine of so large a capacity on so small a budget."

---

[12] Reprinted in Brian Randell (ed.), *The Origins of Digital Computers: Selected Papers* (Springer-Verlag, 1973, ISBN 978-3540113195), pp305-325.

By capacity, Atanasoff literally refers to what the machine would be capable of. At its heart, this was "a computing machine which has been designed principally for the solution of linear algebraic equations," to quote Atanasoff's document, and that meant it could tackle problems that might otherwise take weeks to solve using mechanical aids. He set out nine potential fields of application, ranging from "multiple correlation" to "perturbation theories of mechanics, astronomy and the quantum theory".

He explained: "Since an expert [human] computer requires about eight hours to solve a full set of eight equations in eight unknowns… to solve 20 equations in 20 unknowns should thus require 125 hours. But this calculation does not take into effect the increased labour due to greater chances of error in the larger systems and hence the situation is much worse than this."

This difficulty, wrote Atanasoff, was why the "solution of general systems of linear equations with a number of unknowns greater than ten is not often attempted." His machine would open up a new field of research and possibility. And all for $5330? Who could possibly say no?

Not the Research Corporation, a body founded in 1912 to support scientists and inventors to create patented products that could then be licensed. Its confidence in Atanasoff's work, and the amount of money involved, sparked a sudden interest in the computer from Iowa State College's president, Charles E Friley. After largely ignoring the project, which had been relegated to a basement within one of the Physics buildings, Friley told Atanasoff to sign an agreement that would give the college 90% of the proceeds of future income generated by the computer (and its patents). If Atanasoff didn't agree, Friley threatened, he would withhold the $5330 of funding.

Atanasoff, never afraid to speak his mind, did not bow under this pressure. "I made what I thought was a small noise, but it was regarded by some as a loud one" he said later.[13] After much negotiation, he agreed to hand over the patent to the college but, in return for paying half the cost of the patent, he would benefit from half of any proceeds. It is a reflection of Atanasoff's sense of fairness, and his appreciation of Clifford Berry's work, that he agreed to pay 10% of his income to his assistant.

---

[13] Tammara Burton, *World Changer: Atanasoff and the Computer*, p117

All this happened in July 1941. A month earlier, Atanasoff had played host to John Mauchly, who would go on to co-create ENIAC. The pair first met at the annual meeting for the American Association for the Advancement of Science (AAAS) in late December 1940, after Atanasoff attended a lecture by Mauchly on his use of a harmonic analyser to help forecast weather (as we will discover in the story of the ENIAC, a keen interest of Mauchly's).

Atanasoff approached Mauchly at the end of the lecture and explained his work on the recently finished prototype. They started writing to one another, and it wasn't long before Mauchly asked if he could see the computer in person.

On that fateful visit – Mauchly arrived on the evening of Friday 13 June, leaving early on Wednesday 18 June – the three men (including Berry) would spend hours dissecting the workings of the machine. Exactly what happened would be much discussed and much disagreed about in court three decades later, but from all we know about Atanasoff's personality, and his desire to share information, it would only be natural for him to go into detail with this most eager of students.

His first wife, Lura, felt her husband was being far too open. "Several times during the visit, Lura warned her husband that he was 'talking too much' to this man about whom she had serious misgivings," wrote Atanasoff's granddaughter.[14] "She understood the necessity of protecting intellectual property and sensed in Mauchly a sinister motive."

According to Atanasoff, there was only one request that he denied his guest: to take away a copy of the 35-page report that had been sent off to the Research Corporation. At this point, he was nearing the end of his negotiations with Iowa State College about the patent application, and was wary of sharing any written documentation. He did allow Mauchly to borrow a copy and take notes, however, and Lura believed that their guest stayed up long into the night to copy them verbatim from the report.

Over the next few months, during which Mauchly would continue to write and ask for updates, Atanasoff and Berry dedicated whatever time they could to their computer. It worked, except for a problem with the output. As with so much of the machine and the theory that lay behind it – Atanasoff even invented his own version of Boolean algebra, having been unaware of George Boole's breakthrough

---

[14] Tammara Burton, *World Changer: Atanasoff and the Computer*, p122

in this area almost a century earlier – the two men had few off-the-shelf solutions to their problems.

One was how to output results onto cards that could then be fed back into the machine for the next cycle of operations. They came up with a solution where a 5000V electronic spark scorched a tiny hole in the card to signify the digit one (if it was blank, that meant zero). The cards could then be read because another electronic spark – this time 3000V – could arc through the hole to create a connection. It was yet another ingenious solution to a problem, but not perfect: Atanasoff estimated the failure rate to be between one in 10,000 and one in 100,000. That may sound small, but it's enough to cause inaccuracies over the course of a long calculation.[15]

By spring 1942, this problem aside, the computer was finished. Iowa State College even started a PR campaign, arranging newspaper and radio interviews with the two inventors. 'Machine to Solve Algebraic Problems Replaces 100 Computers', read the headline in numerous papers in June[16] (with 'computers' here referring to humans performing mathematical calculations). It concluded: "The machine is probably the largest ever built. It will replace 100 expert computers with calculating machines when in action. The completion and trial has been delayed by Iowa State College activities in connection with national defence, Atanasoff said. He expects to give the machine an initial trial before the summer."

How frustrating for both men, then, that war intervened before they could find a permanent solution to the output problem. Berry knew he could be drafted at any time: he had now completed his master's degree and the computer was not directly related to the war effort. He chose to take matters into his own hands, accepting a position related to national defence in mid-1942 – and whisking away Atanasoff's secretary, Martha Jean Reed, as his newly wedded wife.

In the scant time Atanasoff wasn't dedicating to additional wartime duties and academic demands, he detailed the documentation necessary for the patent application. An application that Iowa State College never submitted, in part, perhaps, due to the distraction of the war.

---

[15] Alice R Burks and Arthur W Burks, *The First Electronic Computer: The Atanasoff Story*, p63
[16] 'Machine to Solve Algebraic Problems Replaces 100 Computers', Telegraph Press news service, Sunday 14 June 1942

Atanasoff was likewise distracted. For the rest of World War II, he would act as a consultant to the Naval Ordnance Laboratory, moving away from Iowa (and his family, one of the contributing factors to the eventual divorce from his first wife, Lura) to Washington DC. He was rapidly promoted to Chief of the Acoustics Division, earning the Navy Civilian Service Award in 1945.[17]

Still, though, he could not escape the pull of computers. In 1945, he was invited to head up the US Navy's project to "design and build a new high-speed digital electronic computer for general research applications".[18] He accepted, and in February 1946 he first saw a demonstration of the ENIAC.

Neither of the ENIAC's inventors was present at the demonstration, so Atanasoff could ask no detailed questions. Indeed, the computer was then considered to be an object of national security classification, so he could only take any claims at face value. He certainly couldn't peek into the ENIAC's inner workings, as Mauchly had into the ABC so many years previously. What's more, this huge machine was surely light years ahead of the ABC in terms of complexity.

This is why, when Atanasoff found out that the Iowa State College had never paid for his patent to be filed, he was annoyed but not distraught. Times had moved on, and quickly. Plus he was busy with two high-ranking Navy roles, including having to monitor the atomic blasts happening at the Bikini Atoll. Soon after, he was informed that the Navy's computer project was being abandoned.

With Atanasoff still being kept busy with US Navy duties, life continued at Iowa State College. That included a new chair of the Physics Department, who was probably unaware of the now abandoned ABC computer. It's believed that when a grad student asked to use the room, the chair gave his permission to dismantle the machine – it couldn't simply be moved because the chassis was wider than the door. All that is now left of the original machine are the two memory drums and one of the 30 add-subtract modules with its seven dual-triode vacuum tubes.[19]

But for the trial, this would have been the end of the ABC (which at this point, remember, did not even have a name). Its place in computing history was only sealed

---

[17] Tammara Burton, *World Changer: Atanasoff and the Computer*, p134
[18] As above
[19] One drum can be found at the university museum, the other in the Smithsonian, while John Gustafson (who we will hear from shortly) donated the add-subtract module to the Computer History Museum in Mountain View, California, when the ABC replica moved there.

14   The Computers That Made The World

An add/subtract module at the Computer History Museum
Image: Arnold Reinhold, CC BY-SA 4.0

when US District Judge Earl R Larson recorded his verdict on 19 October 1973 that the ENIAC patents were invalid. His third finding was the most controversial, stating: "Eckert and Mauchly did not themselves invent the automatic electronic digital computer, but instead derived that subject matter from one Dr John Vincent Atanasoff." [20]

Effectively, and this remains hugely controversial, the US justice system had named Atanasoff as the creator of the first electronic computer. This would have been huge news, but one day later, President Nixon fired the special prosecutor looking into Watergate, in what would become called the Saturday Night Massacre. There was no space in newspapers to cover the story of a patent dispute.

It took another three months, and the determination of Atanasoff's second wife, to bring it to the attention of an Iowa-based reporter. But it was worth the wait. On 27 January 1974, the *Des Moines Sunday Register* stamped 'COURT: COMPUTER IOWAN'S IDEA' across its front page. This time, the story soon spread across the country, and belatedly the Iowa State College – now named Iowa State University (ISU) – took note. From then onwards, the establishment made much of its part in the history of computing.

This came to a head in 1994, when the ISU decided to build a working replica of the computer. "George Strawn, who had been a recent Department Chair of Computer Science (not the one who disassembled the original ABC) and Delwyn (Del) Bluhm spearheaded the idea and worked to get it funded," said John Gustafson,[21]

---

[20] Judge Earl R Larson's official notes, 19 October 1973.
A copy can be read at **jva.cs.iastate.edu/Legal%20Decision_32-2.pdf**
[21] In email to author

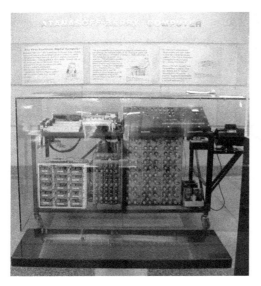

ABC replica at the Durham Center, Iowa State University
Image: Manop, CC BY-SA 3.0

founder of the ISU's Scalable Computing Laboratory and by that point author of several papers on how the ABC operated. On hearing of the project, Gustafson quickly lent his enthusiastic support to it.

"This ABC Replica Project was conceptually important," wrote Del Bluhm, the project's first director, in 2013,[22] "since a working replica of the ABC had to be tested to prove that it did in fact work properly and therefore was the world's first electronic digital computer."

The only problem was that this project would cost hundreds of thousands of dollars. A sum the university was not willing to cover. Fortunately, Strawn and Bluhm secured a large chunk of funding from Charles W Durham, a former student of Atanasoff's at the ISU, and started putting the plan together.

John Atanasoff was still alive at this point, and gave the team access to all his documentation and diagrams. Sadly, he would not live to see the finished replica, dying in June 1995 at the age of 91.

By this time, Bluhm knew the full scale of the challenge. "Our effort involved not only reproducing parts from limited original designs, but also re-engineering designs after interviewing those people who either worked on the original computer or saw the prototype in operation," he wrote.

Finally, by the summer of 1996, an enthusiastic team – including engineers who had worked on early electronics so were familiar with vintage vacuum tubes – were ready to start the building process. The replica was finished in April 1997, but with one caveat: it didn't actually work. While all the components and wiring were present, they weren't necessarily in the right order.

---

[22] Del Bluhm, 'Documenting the ABC Replica Project and its Contributors', in *Atanasoff Today*, Spring 2013, p14

Bluhm's team had done their job of faithfully creating the replica. The challenge was that they were engineers rather than computer scientists, so while all the components and wiring were present, they were still some distance from having the working replica required to prove to naysayers that the ABC computer worked as designed. At this point, John Gustafson was asked to take control of the replica project.

As part of the project, the team had gathered as many people as possible who had seen, heard or worked with the ABC during its short active life, and one of those people was Clara Smith. Previously a secretary in the ISU's Statistics department, she revealed that the ABC had been put to real work by head of the department, Professor George Snedecor.

"She sent [Atanasoff and Berry] systems of equations for least squares fit to be solved," said Gustafson.[23] "They sent back the answers, and then she'd use the desktop calculator, like a Marchant calculator, to check the answer. And that's not much work: that's just about six multiply-adds, and you've got to see whether AX does equal B. But it's much harder to do the inverse – to solve a system with two equations, two unknowns – than to check that it's correct."

Clara Smith explained that Snedecor then used those results in his research papers, which were published at that time. "So I would say the ABC was a production computer," said Gustafson. "It was put into practical use by Snedecor exactly as Atanasoff envisioned."

Due to the nature of human memory, not all of the people Gustafson and his team spoke to gave such detailed answers as Clara Smith, but over time he was able to build up an idea of what the ABC looked like, what it sounded like and tiny details such as where the odometer, which counted upwards to indicate number outputs, was placed. They also discovered that the team building the ABC "had a deck of five cards that was their standard five-by-five test to see whether the machine was operating," said Gustafson.

With all possible information now gathered, it was time to get their hands dirty and start the wiring. Or at least, for Charles Shorb to get his hands dirty. "He was a grad student in my research group who could really do everything," said Gustafson. "He was scared to death at being given the project. 'Frankly,' he said, 'I don't know what to do.' I said, you will."

---

[23] Interview with the author

From the moment he saw the replica, Shorb was in awe of both its original inventors and everyone who had worked on the project so far. "It was the most beautiful wiring I'd ever seen in my life," said Shorb.[24] "You think of old computers taking up a whole room, but [Atanasoff] created basically an automatic Marchant calculator that's the size of a desk. That's phenomenal."

The ABC was so compact because no space was wasted. "The control circuits were sat right behind the add/subtract modules, then you have the circuits that sat behind that control the thyratron firer for the base two input/output," said Shorb. "And then right behind that were the control circuits where the wiring went up to the commutators that actually was the brains of the whole thing." Looking at this, following the trail of wires, was reminiscent of modern circuit design, Shorb added.

He then almost ruined the whole project by switching it on, a move greeted with wisps of smoke. Fortunately, that problem was easy to debug – the power bus was wired incorrectly – and he set to work. Painful work, as it turns out, because the extreme voltage, going from −120V to +120V, resulted in "way too many" electric shocks for the young grad student.

Despite the pain, what followed was three months of working "day and night" on the wiring and the circuitry, to ensure that the replica worked exactly as Atanasoff and Berry intended. Fortunately, they had left behind detailed schematics. Despite his work ethic and evident motivation, Shorb could not finish everything before he left in September to join Intel. Due credit should go to Guy Helmer, now a professor at ISU, who later completed the wiring.

By the time Shorb left for Intel, he had wired five of the fifty blocks, and that was enough for the team to achieve its goal: to create a working replica. One that would prove wrong any doubters – the ABC was more than just a concept.[25]

So what was it like to use? While in operation it was "no louder than a sewing machine, said Gustafson, "from an electrical signal viewpoint it was horrifically noisy". "There was so much jitter. Brushes on metal produce a terrible oscilloscope signal, let me tell you, and I thought this machine was never going to work. But the extreme voltage was enough that even a noisy signal was able to compute and make

---

[24] Interview with the author
[25] While the public can't try the ABC out for themselves, visitors to the Iowa State University computer lab can still admire the replica. You can also see it in action on YouTube thanks to a 1999 recording featuring Gustafson and Shorb at **youtu.be/YyxGIbtMS9E**

things click along. And so it was a thrill when we gave it a problem and it actually solved it."

It was time to go public, which the team did at an event at the National Press Club in Washington DC on Wednesday 8 October 1997. Gustafson and Shorb devised a simple problem where the answer was three: 9 - (3×2) = x. They explained to the watching journalists – which included staff from the National Geographic and Time Magazine – that when they heard three clicks on the odometer they knew the computer had worked its magic.

"Switches were flipped, a punch card was inserted and an electric motor turned a drum studded with copper contacts," wrote Kenneth Pins in the write-up for *The Des Moines Register*.[26] "It was 2½ minutes later when the numbers on a device resembling a card's odometer advanced three clicks, revealing that the answer was 3."

After heading out on a nationwide tour, the ABC replica now resides at the Computer History Museum in California, next to the ENIAC. As one of the few people to operate the ABC replica, we asked Shorb if he thought the ABC's impact would have been different if Atanasoff and Berry had the time to solve the card output problem. "Absolutely," he said, adding: "I think that history would have been more kind to the ABC if they would have gotten the patent."

But he also points out something that we hope will become clear through the other stories in this book. "When you really take a look at the history of computing as a whole, you realise that no single person can claim anything… but you at least should give credit where credit is due for the advances that people make."

Dr John Atanasoff and Clifford Berry certainly deserve credit, as do the team who built the ABC replica almost 60 years later.

---

[26] Kenneth Pins, 'Computer age dawns again thanks to ISU', in *The Des Moines Register*, Thursday 9 October 1997, p1

Zuse Z3 replica on display at Deutsches Museum in Munich

Image: Venusianer, CC BY-SA 3.0

Zuse Z3

**From toy construction sets
to building a pioneering
electromechanical computer**

When Konrad Zuse's parents gave their young son a metal toy construction set – much like the Meccano sets popular in Britain and the USA at the time – they had no idea that they were sowing the seed for the world's first binary, programmable computer. For Zuse, with help from family rather than companies, friends rather than universities, created a series of machines that could rival any of their contemporaries.

If he had been based in Boston rather than Berlin, there's every chance that his most admired computer, the Z3, would be as famous as the ENIAC. And that Konrad Zuse, the man, would be as well-known as John Mauchly in the US and Alan Turing in the UK.

For now, though, let us spare a thought for his long-suffering parents, who not only contributed money to his first computer but would eventually donate their dining room to house it too.

Arguably, this was their own fault. That first construction set led to Konrad spending so much time creating mechanical inventions that he won prizes for them – prizes of extra construction parts that he would turn into ever-more elaborate models. He was particularly proud of a large grab crane that sat on top of the wardrobe in his room and could be controlled using strings from his bed.

So formed a pattern – arguably an obsession – of solving problems using mechanical devices. Zuse devised numerous ways to automate the world, many of which he shared in his autobiography, *The Computer – My Life* (*Der Computer – Mein Lebenswerk* in the original German). He even built an automatic fruit-dispensing machine, much to the delight of friends such as Andreas Grohmann who described this "mandarin machine which took money and gave mandarins, and sometimes indeed returned the money with the goods".[1]

With a mind that refused to run along conventional lines, Zuse did not follow a straightforward academic path. Equally gifted at mathematics and drawing, it wasn't clear to the teenage Zuse what his career would be. Artist? City planner? Photographer, even? But in the end, engineering would win.

This led him to Berlin's Technical College, which meant that a 17-year-old Zuse had to leave his parents' home in Hoyerswerda – roughly 100 miles south of Berlin – and rent a small room on his own. Fortunately, he quickly forged new friendships via the Akademischer Verein Motiv, a student club with a theatrical side. The Motiv,

---

[1] Konrad Zuse, *The Computer – My Life* (Springer-Verlag, 1991, ISBN 978-0387564531), p35

and the Motivers who made up its membership, would prove crucial not only in Zuse's personal life but also for becoming the loyal group of friends who would donate their time and their money (even their parents' money) in support of his computers.

Academically, progress wasn't so smooth. Zuse soon regretted choosing mechanical engineering as his major, complaining that the "creative spirit was left little freedom in the manner of presentation; everything was standardised, everything was decided".[2] He switched to architecture, but "Doric and Ionic columns left me cold", which led him to his final choice of civil engineering.

Konrad Zuse
Image: Wolfgang Hunscher, CC BY-SA 3.0

It was here Zuse met his nemesis: statics. This branch of applied physics concerns static bodies such as bridges and roofs, calculating the forces in play upon each of their interconnected parts. Zuse admitted a "pronounced aversion" to the science, adding that he "worshipped the professors in command of this calculus as if they were demigods from another world".[3]

As a student, Zuse faced such problems on a regular basis. Most of the time, according to renowned computer historian Paul Ceruzzi,[4] he would solve them using a slide rule, but he also had experience with mechanical desktop calculators. The challenge wasn't each calculation but "transferring the results of one operation to another part of the problem",[5] wrote Ceruzzi.

The increasingly sophisticated engineering designs of the early 20th century only compounded the challenge. While an individual could handle a series of six equations with six unknowns, real-world projects were far more complex. Ceruzzi cites an example of calculating the stresses on a roof that required a system of 30 equations with 30 unknowns; this took a team of human computers months to work out.[6]

---

[2] Konrad Zuse, *The Computer – My Life*, p15
[3] As above, p16
[4] Paul E Ceruzzi, 'The Early Computers of Konrad Zuse, 1935 to 1945', in *IEEE Annals of the History of Computing*, Vol 3, No 3, July 1981
[5] As above, p242
[6] Paul E Ceruzzi, *Reckoners* (Greenwood Press, 1983, ISBN 978-0313233821), p11

In typical style, Zuse sketched out mechanical methods to simplify the problem while still a student. One involved a crane, reminiscent of his schoolboy invention, that would automatically "scan the plane of the machine and enter the values into their correct places". He then realised that he could turn this physical design into something more abstract, replacing the crane with the concept of a 'selector' and the numeric values into what we would now call memory. There are clear parallels with the store and reader of Babbage's analytical engine, of which Zuse was (at that point) entirely unaware.

He captured the evolution of his thoughts and designs in his diaries, but we should put aside any notion that Zuse ever followed a straight line. In his autobiography, he describes his "many detours and by-ways"[7] during his university days, which included a year away from studying where he tried to earn a living designing adverts. He finally graduated with a diploma in 1935, eight years after starting at the Technical College. His parents must have breathed a sigh of relief when, soon after, he landed a job as a structural engineer at the Henschel Aircraft Company. Surely their mercurial son would at last settle down to a normal life.

This was not to be. His work called for yet more calculations, fuelling Zuse's desire to delegate such tedious work to a machine. By this time his parents had moved to an apartment in Berlin, and this was where – after quitting his full-time job at Henschel, although he would continue to work for them part-time – he "set up an inventor's workshop"[8] to turn his sketches and diagrams into reality.

"One day, shortly after he had received his degree and had been working as a structural engineer but for a few months, he explained to us, a few of his pals, that he was planning to build a universal[9] computing machine," wrote Andreas Grohmann, a fellow Motiver who had just qualified, at the age of 20, to be a mining engineer. "He was looking for helpers… I agreed."

Grohmann describes his "months working all day long with Zuse" to build the machine, across the summer and autumn of 1936 but primarily over the summer of 1937. Zuse's parents' sacrifice yet again comes into clear view through Grohmann's account: "He had set up a small workshop in a small room in his parents' apartment

---

[7] Konrad Zuse, *The Computer – My Life*, p21
[8] As above, p33
[9] Andreas Grohmann is using some poetic licence here as the Z1 would not be considered as a 'universal' computing machine due to its lack of support for conditional branching.

on Methfesselstraße in Berlin, and also used the large living room – now off-limits to his family – as a construction space for his machine."

Over the next two years, during evenings and weekends, Zuse would work tirelessly to create his first machine: what would retroactively be called the Z1. "He was really obsessed with his work," says computer scientist and historian Raúl Rojas.[10] "I think that this obsession is what explains how he could do so many things in 24 months, from 1936 to 1938. He had essentially finished [the Z1] before the war broke out."

Rojas argues that Zuse "founded the first computing startup". Every other computer covered in this book had the backing of a university, corporation, or government – sometimes a mix of all three – whereas Zuse financed his first computer through the generosity of his friends and his family. His father, who had retired from the Post Office, even returned to work for a year to raise funds for materials.

Zuse used his persuasive charm on friends such as Grohmann to donate both money and time. Rolf Edgar Pollems, another Motiver, wrote that "we younger students would have gladly helped him realise his dream of building an operational machine… we sawed out metal shapes with the fretsaw according to his instructions, and carried out other similar tasks to help him."[11]

This despite the fact that the theory behind the Z1's operation was well beyond their experience or understanding. "I'm honest enough to admit that I worked blind," wrote Grohmann. "I did not know precisely how this monster, that was taking shape there, was supposed to function. And yet, the machine was completed, worked with an unholy rattle, and supplied precise solutions to complicated tasks."

Indeed, it would have been beyond the ken of anyone who considered themselves an expert at the time, as Zuse took an unconventional approach to his machine. Until then, most calculators operated on the simple principle of repeated addition: if you wanted to multiply seven by six, you would start off with six in the machine's register and then repeat the addition process seven times.[12] By contrast, because Zuse's arithmetic unit worked in binary rather than decimal, multiplication becomes as

---

[10] Interview with author. Raúl Rojas has written the definitive book on the Z1, Z2, Z3, and Z4 architectures, *Konrad Zuse's Early Computers: The Quest for the Computer in Germany* (Springer, October 2023, ISBN 978-3031398759)

[11] Karl-Heinz Czauderna, *Konrad Zuse, der Weg zu seinem Computer Z3*, p85 (translation from German found in Konrad Zuse, *The Computer – My Life*, Springer-Verlag, p36)

[12] There were more sophisticated methods around, including one where the sum was split into parts, but at this stage repeated addition (and subtraction) were the dominant and most cost-effective approaches.

simple as addition (in binary 1 × 1 = 1, 1 × 0 = 0, 0 × 0 = 0... while 1 + 1 = 10, 0 + 1 = 1, 0 + 0 = 0).

"The most astonishing thing for someone who studies what Zuse did is that he rediscovered so many things," says Rojas. Zuse had no mathematical training, was ignorant of academic developments in that field, so it was his engineer's instinct that pushed him to use binary rather than decimal, to devise his own square root algorithm (introduced in the Z3) and even his own logic.[13]

"Floating-point had already been thought of for calculators and he rediscovered that," says Rojas, "and he also discovered the algorithms for dealing with floating-point numbers. And that's the most amazing thing for me." Using floating-point numbers meant the Z1 and its successors could reduce complex real numbers such as 143.75 into component parts: 143.75 would become 1.4375×10^2.

Where Zuse's computers stand head and shoulders above early contemporaries such as the Harvard Mark I is the theory that underlies them. Remember, Zuse built his computers in near-isolation: he devised his own form of calculus, accidentally replicating the work of mathematicians. The only theory he drew upon, to the best of our knowledge, is Leibniz's work on binary mathematics.

The Z1 was a purely mechanical affair, but if you're imagining something akin to a typewriter then think again. This was a sprawling machine, with sliding metal rods to represent the zero or one of each 'bit' of memory, that consumed roughly the same amount of space as three typical dining tables placed side by side. In certain areas, it also stood three feet high.

If you can't visit the reconstruction of the Z1 at the Deutsches Technikmuseum in Berlin, which Konrad Zuse himself oversaw in the 1980s, then the next best thing is to watch the YouTube video.[14] This also does an excellent job of showing the sophistication of the Z1's design, complete with its separated memory store and arithmetic control units. When working smoothly, this could perform addition or subtraction in around two seconds, with division and multiplication taking a few seconds longer. Operators could enter different functions using a film strip punched with holes.

---

[13] Zuse called his logic Conditional Combinatoric (Bedingungskombinatorik), which he created to describe the operations his computer would process. This tied in (not that Zuse knew this) with a movement in mathematics to reduce all operations to base logical expressions (work done by David Hilbert and W Ackermann in *Foundation of Logic*, published in 1928, and separately by Alfred North Whitehead and Bertrand Russell in *Principia Mathematica*).

[14] See it at youtu.be/HDxs3-aJSAI

The big caveat there is 'when working smoothly'. While Zuse's friends did their best to follow the precise measurements for each metal part, they were still cutting by hand. And even the reconstruction, with its precision-cut components, proved problematic. Zuse, in typically self-deprecating style, wrote: "[The Z1] didn't work well at the time and the replica is also very faithful in this respect: it doesn't work well either."[15] Zuse could determine and correct the replica's faults, but since his death in 1995 no-one has been able to fulfil that role.

So, although a work of genius, the Z1 was not a practical device. Improvements were clearly necessary, so Zuse set to work on what has come to be known as the Z2. This moniker is a little misleading, as the Z2's main role was to act as a proof of concept that showed Zuse could replace the currently mechanical arithmetic unit with one built using electromagnetic relays. It was an idea he developed with his friend, Helmut Schreyer, who was studying telecoms engineering.

Why not vacuum tubes? These were suggested by Schreyer, but the pair were discouraged from taking this approach when they presented a prototype of how a vacuum tube switch might work to professors at the Technical University in Berlin. When Zuse and Schreyer explained that a complete computer would need 2000 vacuum tubes to work, they were met with shakes of the head. Too expensive, too unreliable.

Schreyer would continue his research into the vacuum-tube relay circuits, earning a PhD in Engineering in 1941 as a result, but it would be many years before Konrad Zuse's computers would incorporate them. In 1942, Schreyer even proposed creating a version of the Z3 with vacuum tubes to the German Army command, but they turned the idea down when Schreyer explained it would take two or three years to build such a computer. By this time, he was told, the war would be won.

Schreyer was a member of the Nazi party, but Zuse never joined. In the early 1930s, he had spent a few weeks of basic army training with the Reichswehr, the Reich Defence, but more as a preventative measure: others who had not shown loyalty to Hitler and the Third Reich were "being forced into line and marched

---

[15] This is a translation from the original German, which is: "Dieses Modell steht als Nachbau im Museum für Verkehr und Technik in Berlin. Damals hat es nicht gut funktioniert und auch der Nachbau ist insofern sehr getreu, auch er arbeitet nicht gut." From *Geschichten der Informatik: Visionen, Paradigmen, Leitmotive* (Springer, 2004, ISBN 978-3540002178), page 36

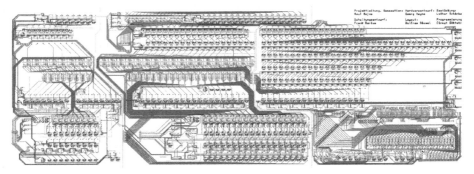

Layout diagram of the Z3 processor
Image: courtesy of Raúl Rojas

off",[16] so he and eleven other classmates thought it better to volunteer than be commandeered.

Zuse was drafted into the army proper when Britain declared war in September 1939, after Hitler had annexed Czechoslovakia. By this time, he had almost finished the Z2, and though his patron Kurt Pannke – who ran a company that made specialist calculators and had given Zuse 7000 Reichsmarks to help him build the Z2 – wrote a letter to the authorities asking for Zuse to be granted leave so he could continue his work, this met with no success. It didn't help that one of Pannke's arguments was that the Z2 had applications in aircraft construction. On seeing the letter, the battalion commander summoned Zuse and asked him to explain himself: "The German Luftwaffe is top-notch; what needs to be calculated there?"[17]

Zuse wisely declined to criticise the Luftwaffe's planes and his permission for leave was refused. Fortunately, he found a more sympathetic audience in his captain, who would sometimes allow the young inventor the chance to work on his ideas in the quiet of the captain's room. Nor did Zuse face any combat action, using his six months in the infantry to work on his ideas and to "formalise the rules of chess using mathematical logic".[18]

---

[16] Konrad Zuse, *The Computer – My Life*, p30
[17] As above, p57
[18] As above, p57

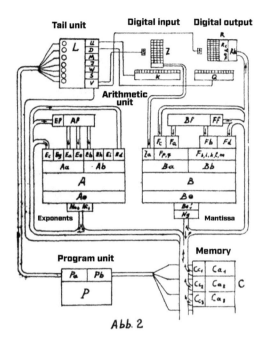

Z3 architecture sketch from the 1941/1950 patent application

His time in the army ended when Professor Herbert Wagner, who was developing remote-control flying bombs at the Henschel Aircraft Company, recruited Zuse as a structural engineer. By all accounts Wagner was a brilliant man, and one of many German scientists who migrated to the USA immediately after the war to help their military efforts.[19]

Now back in Berlin, and with evenings and weekends free, Zuse could continue to develop the Z2 whilst working for Wagner. By September 1940, the computer was ready for its first official demonstration, to Professor Teichmann of the German Aviation Research Institute (the Deutsche Versuchsanstalt für Luftfahrt, shortened to DVL). "Just hours before then I was still frantically trying to get the machine to run, but it was useless," wrote Zuse. "The guests arrived – I was sweating blood – and behold, my machine performed flawlessly."[20]

This demonstration proved enough to persuade Teichmann and the DVL to sign a contract with Zuse to complete work on the Z3 – from an architecture point of view, the same as the Z1, except now with a square root algorithm. Rojas describes this algorithm as the "jewel in the crown of the Z3",[21] avoiding the traditional approach

---

[19] Zuse was one of several people who helped Wagner develop the Hs 293 guided missile. Although Zuse downplays the missile's success in his autobiography, saying the Allies learned to jam the radio control systems, this only happened after it had sunk several ships. But as is also clear from the book, Zuse did not appreciate the horrifying extent of the Nazi regime, with rumours of concentration camps being easily dismissed as foreign propaganda. It seems likely that it was only as the war approached its end, when Zuse saw emaciated workers for himself, that he fully appreciated what had been going on.

[20] Konrad Zuse, *The Computer – My Life*, p61

[21] Raúl Rojas, 'Konrad Zuse's legacy: the architecture of the Z1 and Z3', in *IEEE Annals of the History of Computing*, 1997, Vol 19, Issue 2, p13

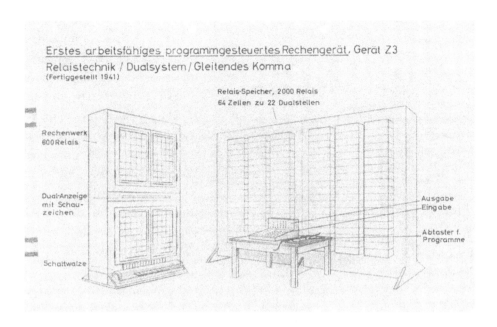

First functional program-controlled calculating device, device Z3 relay technology / dual system / floating point (Completed 1941); sketch from 1941/1950 patent application

- **Rechenwerk 600 Relais** – Arithmetic unit, 600 relays
- **Dual-Anzeige mit Schauzeichen** – Dual display with indicator
- **Schaltwalze** – Shift drum (impulse generator)
- **Relais-Speicher, 2000 Relais 64 Zellen zu 22 Dualstellen** – Relay memory, 2000 relays, 64 cells (i.e. words) of 22 dual positions (i.e. bits)
- **Ausgabe** – Output
- **Eingabe** – Input
- **Abtaster f. Programme** – Scanner (punched tape reader) for programs

of guess-and-guess-again until you don't get an error. Instead, the calculation stepped through a precise 18 cycles to determine the square root of any number.

But the chief difference between the Z1 and the Z3 was the use of electromagnetic relays to represent zeroes and ones rather than clunky metal plates. Astonishingly, Zuse completed it before the end of 1941, and at that point the Z3 was arguably the most technologically advanced computer in the world.

Not that you would know this to look at it. Assembled from second-hand materials of various provenances, the relay coils needed much fine-tuning to ensure they activated at the same time. It's one of the reasons why Zuse was the only person who could service the machine when it was complete. Despite this, the Z3 was "relatively reliable", Zuse wrote,[22] although we should bear in mind that the relative in question was the Z1.

Like the Mark I, the Z3 could calculate with the minuscule and huge numbers necessary for scientific work. It was certainly more than capable of coping with calculations associated with statics calculations in aircraft manufacture, which is what the DVL was interested in.

In his 1997 paper,[23] 'Konrad Zuse's Legacy: The Architecture of the Z1 and Z3', Rojas goes into detail on how this worked. In particular, he explains how "the smallest number representable in the memory of the Z3 is $2^{-63} = 1.08 \times 10^{-19}$, and the largest is $1.999 \times 2^{62} = 9.2 \times 10^{18}$". That's an enormous range.

Still, though, the Z3 had obvious limitations: not due to the size of the numbers it could store but to how many it could store. This was primarily due to the scarcity, price, power demands, and size of electromagnetic relays. Unlike Howard Aiken when building the Harvard Mark I, with the luxury of IBM's workshops and financial backing, Zuse had to pay for everything out of the scant money provided by the DVL and his own backers. Using Zuse's notes from the time, Ceruzzi estimated the cost at around $2 per relay, which was a huge sum in 1940s Germany.

The Z3's storage was split across two cabinets, with each holding 32 columns of 28 relays (896 relays in total). Because six of the relays per column were needed for operation, this left 22 relays per column, each of which could represent a one or a zero. So one cabinet could hold 32 numbers represented by 22 zeros and ones (we would now call these 'bits').

While 64 numbers sounds like a lot, it wasn't enough for the complex calculations required by the DVL. To tackle these, the computer needed to hold hundreds of variables within its storage at any one time, and so while the Z3 worked as an excellent proof of concept it wasn't the real deal. While Zuse could show that it was capable of tackling the type of problems the DVL was interested in, such as calculating the determinant of matrices, they were simplified versions that could already be solved efficiently by humans.

---

[22] Konrad Zuse, *The Computer – My Life*, p63
[23] Raúl Rojas, 'Konrad Zuse's legacy: the architecture of the Z1 and Z3', in *IEEE Annals of the History of Computing*, 1997, Vol 19, Issue 2, p7

Still, it's worth admiring what Zuse had created and what it looked like. To start with, equations could be fed into the Z3 using punched tape. This was standard celluloid film, as used in movie theatres, but the reader that scanned in the instructions (zeros and ones) was custom-made. This equation was then sent to the arithmetic unit, so that it knew which calculations it needed to perform.

The next step was to load the equation's variables into the memory banks (the 64 columns mentioned above). To do this, operators used a keyboard to enter the numbers as decimals. The Z3 handled conversion from decimal to binary, but they needed to be entered as four digits to the power of ten (or minus ten). They were accurate to four decimal places and ranged from 0.000000001 (a nanometre, or 10^{-9}) to 1,000,000,000 (one billion or 10^9). Once the program started running, new numbers would be sent to available, empty storage banks.

Just like modern computers, the Z3 ran on a cycle. A smartphone made in 2025 typically includes a processor that runs at around 2.5GHz, which is 2.5 billion cycles per second. By contrast, the Z3 ran at around five cycles per second. According to Raúl Rojas, addition took three cycles on the Z3, multiplication took 16 cycles, division 18 cycles, and a square root 20 cycles. That translates into roughly a second, three seconds, four seconds, and five seconds respectively.

After an appropriate wait – the programs would involve a complex series of such calculations – the answer would appear on the display. Like the input, this would be in decimal, and appear as a series of four digits (via labelled bulbs) to the power of 10^{-12} to 10^{12}. Looking at a photo of the reconstructed Z3, it's striking how much more compact it is than the Mark I or ENIAC.

The replica only includes one storage unit whereas the actual Z3 included two, each filling a cabinet roughly the size of a wardrobe. The other unit contains the arithmetical unit – consisting of around 600 relays – the tape reader, and the control unit, which needed roughly 200 relays to convert punched instructions into code.

It's an incredibly impressive technical feat, and a replica is normally on display as part of the Computers Exhibition at the Deutsches Museum in Munich (the exhibition is closed due to refurbishment at the time of writing, but due to reopen in 2028). You can also find a replica at the Konrad Zuse Museum in Hünfeld.

Detractors of the Z3 will point out that it lacks two key features: the ability to store programs and conditional branching. The former is inarguable – but hardly surprising – while the lack of conditional branching is primarily due to the lack of relays available to Zuse. He was certainly aware of the theoretical value of creating programs that could work on the basis of 'if this happens then do that'.

The Z3 was effectively a proof of concept, something to demonstrate the power of a programmable, electromechanical binary computer, and it served this purpose well. For example, records exist from a visiting group of mathematicians who were suitably impressed when it computed a determinant on demand.

The DVL would also benefit from its development, with Zuse creating a special-purpose version of the Z3 – called the S1 – that was used to calculate the deviations of cheaply produced wings on the Henschel Hs 293 flying bomb. According to Ceruzzi, when completed in 1942 the S1 "replaced a staff of 30 to 35 women who had worked around the clock with electric calculating machines".[24]

By this time Zuse had assembled a team of people to work with him, and they set to work building the Z4 for the German military. It was many times bigger than the Z3 due to its far increased mechanical memory capacity, but the biggest hindrance proved to be the increasingly frequent and more intense Allied air raids on Berlin. Zuse had to switch his workshop's locations several times.

Zuse and his team were close to completing construction of the Z4 by early 1945. However, the authorities decided it should be moved away from its increasingly perilous position in Berlin; Allied bomb attacks had already damaged the workshops where construction was taking place, and the Z3 had been destroyed in an air raid the previous year. They initially sent the Z4, and Zuse, to the safer university city of Göttingen, where he and his team managed to complete construction and run test programs.

"This was the moment for which I had waited for ten years – when my work finally brought the success I desired," wrote Zuse.[25] "It was now very tragic that precisely in these days the Americans stood ready in [the nearby town] Kassel."

So it was time to move on once more, this time to Hinterstein, a picturesque village tucked in the Bavarian Alps. For the next three years, the Z4 would sit largely

---

[24] Paul E Ceruzzi, 'The Early Computers of Konrad Zuse, 1935 to 1945', in *IEEE Annals of the History of Computing*, Vol 3, No 3, July 1981, p256
[25] As above, p57

forgotten in a cellar. But not entirely forgotten. In 1947, as part of a team of Allied investigators, American mathematician Roger Lyndon interviewed Konrad Zuse to discover more about this rumoured German computer. He also saw the machine, but not in action.

Zuse was not inclined to divulge all his secrets to investigators, having had his computers, workshops, and his apartment – along with his parents' apartment – damaged and even destroyed by Allied bombers. And with severe restrictions on post-war Germany in terms of scientific research and development, it's unlikely he was keen to share his full ideas with American visitors.

Still, it's interesting to read Lyndon's assessment of the computer in his simply titled 1947 article, 'The Zuse Computer'.[26] While he sticks mainly to a prosaic description of the computer and its abilities, he is critical of its "rather rough homemade components" and says an "important limitation upon programming is that the machine must adhere to a prescribed linear course of operation". In other words, no conditional branching.

Lyndon also provides an insight into the conditions Zuse was working under, which he describes as "rather primitive", adding that Zuse is working "with inadequate material". He adds that Zuse only has one assistant to help him.

Even if Zuse's hands were tied from a physical viewpoint, his mind was busy. During the war, he had begun work on his own programming language, and this would evolve into what he called Plankalkül – 'plan calculus' in a direct English translation, but perhaps better understood as a 'calculus for computing plans'. This wasn't about numbers but about logical problems; Zuse was trying to formalise a way to tackle them using what we would now call a programming language. For instance, he set out – in enormous detail – how you would use Plankalkül to create a chess program.

While modern-day programmers would not recognise the Plankalkül's syntax, it set out concepts that are now commonplace – particularly variables and subroutines. Zuse's original intention was to turn this into a thesis for a PhD, but he instead wrote an unpublished book on the subject, which he finished in 1945. Three years later, Plankalkül made its public debut in a paper Zuse presented to the Annual Meeting of

---

[26] Roger C Lyndon, 'The Zuse Computer, Mathematical Tables and other Aids to Computation', in *The National Research Council*, Vol 2, No 20, October 1947, pp355-359

The Z3 rebuilt in 2001 for the 60th anniversary of the patent application
Image: courtesy of Georg Heyne

the German Society of Applied Mathematics and Mechanics in 1948, and although a paper was subsequently published it failed to garner any attention.

This would irritate Zuse in later life, but better news was around the corner thanks to a visit from Professor Eduard Stiefel of the Swiss Federal Institute of Technology (in German, die Eidgenössische Technische Hochschule Zürich, which we'll shorten to ETH). Stiefel had just returned from a tour of America's computers, in search of technology to replace the desktop calculators the ETH currently relied on. Zuse quotes Stiefel as saying he had seen "many beautiful machines in beautiful cabinets with chrome work,"[27] with the latter no doubt including the Harvard Mark I with its $50,000 cover.

Stiefel had heard of a German computer tucked away in a mountain village, and must have been thoroughly shocked when he saw the Z4 in a state of disrepair hiding in a cellar. But it worked. Zuse wrote that Stiefel "dictated a simple differential equation, which I was able to program immediately, feed into the computer, and

---

[27] Konrad Zuse, *The Computer – My Life*, p118

solve".[28] This, Zuse reports Stiefel as saying, was the first time that anyone had been able to complete the problem so quickly.

According to Ambros Speiser, however, it wasn't Stiefel who had physically seen the American computers but his two assistants, Speiser himself and Heinz Rutishauser. In the book *The First Computers: History and Architectures*, Speiser goes into great detail on how the Z4 came to arrive at the ETH.[29]

The pair received a letter from Stiefel describing the Z4 and asking them to gain Professor Howard Aiken's view of the computer. "Aiken's reply was very critical – the future belongs to electronics and rather than spending time on a relay calculator, we should now concentrate our efforts on building a computer of our own," wrote Speiser. "We reported Aiken's opinion to Stiefel stating, however, that we did not fully agree with him and that, in our opinion, the proposition should certainly not be flatly rejected."

Stiefel, being in the business of solving mathematical problems, agreed. And he literally decided to put parallel computing into action: the ETH would pay Zuse 50,000 Swiss francs up front and effectively rent it for five years. And they would have the option to pay a further 20,000 Swiss francs at the end of that period to buy the computer outright. At the time, these were enormous sums.

The parallel computer was the ETH's own electronic machine, which Stiefel knew would take a few years to build. This would become the ERMETH, Switzerland's first electronic computer, built upon similar lines to the EDVAC and EDSAC (therefore using von Neumann architecture). According to Hans Rudolf Schwarz's account in the *IEEE Annals of the History of Computing*,[30] it ran from 1956 to 1964.

This means the Z4 was the ETH's main computer for around six years, and it was put to excellent use. Schwarz describes how it checked the security of a dam, the deformation of a Swiss railway bridge, and was used in the design of the first Swiss military plane that reached supersonic speeds.

---

[28] Konrad Zuse, *The Computer – My Life*, p118
[29] Ambros P Speiser, 'Konrad Zuse's Z4: Architecture, Programming, and Modifications at the ETH Zurich', in Raúl Rojas and Ulf Hashagen (eds.), *The First Computers: History and Architectures* (The MIT Press, 2000, ISBN 978-0262181976), pp263-276
[30] H R Schwarz, 'The Early Years of Computing in Switzerland', in *IEEE Annals of the History of Computing*, Vol 3, Issue 2, April 1981, pp121-132

When Stiefel had seen the Z4 in Zuse's cellar, it was functioning but not complete. Part of the contract between Zuse and the ETH was to both finish the machine and add capabilities that the university would find useful, most notably conditional branching, and this meant it wasn't installed in Zurich until August 1950.

These funds also allowed Zuse to invest in his own computer company, which would go on to build numerous relay computers until the mid-1950s before moving on to electronic computers. Due to cashflow problems his company was eventually bought by Siemens, ending Zuse's entrepreneurial career but by no means his interest in computing. He is arguably the father of the idea of 'digital physics', a *Matrix*-style concept that suggests the universe could simply be a computer program.

So where does Konrad Zuse, and the Z3 computer in particular, stand in the pantheon of creators? "Intellectually, I wouldn't put him up there with John von Neumann and Alan Turing," says Paul Ceruzzi,[31] who interviewed Zuse in the 1980s and has written almost 20 books on the history of computing. However, Ceruzzi emphasises that Zuse deserves credit for the logical design of the Z3. "In the early days, most people designed [computers] on an ad-hoc basis, and didn't have much theory. He had theory, which I think put him way ahead of other people, although he never got credit for that."

Raúl Rojas agrees, and puts forward a reason why. "The problem with Zuse is that he did astonishing things, but nobody knew what he was doing outside of a small circle in Germany because of the war. He was not communicating with anyone, not even outside of Berlin." Compare this, says Rojas, with his British counterpart. "[Turing] wrote his papers, he had immediate success, he became very well-known and he was also building computers himself. Zuse was rediscovered [outside Germany] in the 1950s and that was too late for him to have a big impact."

31 Interview with author

Bell Labs Complex Number
Calculator (Model I) relay
and switch frame

Image: © Nokia Corporation
and AT&T Archives

Complex
Number Calculator

"An organisation of intelligent men"
Frank Jewett, Bell Labs

To understand how the Complex Number Calculator[1] came into being, you need to understand the role of Bell Labs in 1930s America. And to understand that, you need to jump in a time machine and travel back to 1909 San Francisco. This was three years after the earthquake that had destroyed swathes of the city. Three men looked on: Frank Jewett, John Carty, and Theodore Vail.

Writer Jon Gertner vividly describes Jewett as "a slight, balding, cigar-smoking physicist" in his celebratory book about Bell Labs

Frank Jewett
Image: Underwood and Underwood, New York – MIT Museum, Public Domain

called *The Idea Factory*.[2] Jewett had been lured from a teaching position at MIT to join the company five years previously, and had quickly risen up the ranks. Carty was the company's chief engineer, while Vail had become president two years earlier in 1907.

Gertner describes Vail as "rotund and jowly, with a white walrus moustache, round spectacles, and a sweep of silver hair", but most importantly as thoughtful. Before he became its president, the American Telephone and Telegraph Company (AT&T) was known for its aggressive policies that often landed it in legal trouble; Vail realised the company would achieve its aims more profitably if it geared its efforts towards technological leadership.

The three men were in San Francisco to discuss how they could achieve exactly this, and gain some rather nice publicity in the process, by setting up the country's first cross-continental telephone link. One that would span the 2500 miles (4000 kilometres) between San Francisco and the AT&T headquarters in New York. They had five years to make it happen, with the Panama-Pacific International Exposition (better known as the San Francisco Fair) due to take place in 1914.

---

[1] Also known as the 'Complex Number Computer', 'Complex Computer', and (later) 'Model I'.
[2] Jon Gertner, *The Idea Factory: Bell Labs and the Great Age of American Innovation* (Penguin Books, 2013, ISBN 978-0143122791)

Carty and Jewett promised Vail they could perform this miracle in time, but rather than rely on their existing engineers they decided to bring in expertise from the academic world. Jewett knew just who to turn to: his old friend Robert Millikan, who he had boarded with when they were both at the University of Chicago.

In his autobiography,[3] Millikan remembers his conversation with Jewett in detail. He quotes Jewett asking for help in one specific way: "Let us have one or two, or even three, of the best of the young men who are taking their doctorates with you and are intimately familiar with your field." His field being electron research. "Let us take them into our laboratory in New York and assign them the sole task of developing the telephone repeater."

Teletype operator using the Complex Number Calculator
Image: © Nokia Corporation and AT&T Archives

One particular student, Harold Arnold, had particularly impressed Millikan through his experimental work, and in early 1911 – having completed his PhD – the 27-year-old travelled to New York. He spent the next two years tackling the problem, which boiled down to creating a high-quality amplifier that could act as a repeater across the thousands of miles separating America's west and east coasts.

The key challenge facing Arnold is that when an electric signal is sent down even the best copper wire, its signal drops off. The existing generation of repeaters, mechanical amplifiers, could only work a handful of times before they distorted the sound of the human voice beyond recognition. For a cross-continental call, they needed better technology.

Arnold developed a "mercury arc telephone repeater which worked reasonably well", according to Millikan,[4] but also improved upon an existing invention of Lee

---

[3] *The Autobiography of Robert A Millikan* (Macdonald & Co, 1951), p134
[4] As above, p135

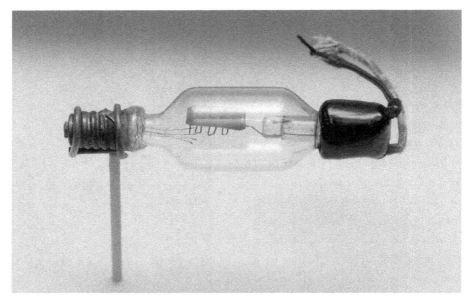

The Audion vacuum tube could be used as a repeater for long-distance communication
Image: Gregory F Maxwell, taken at The History of Audio: The Engineering of Sound, GFDL 1.2

de Forest in 1906. Called the Audion, this early vacuum tube was intended to receive "wireless Morse dots and dashes," said Millikan, but Arnold realised that it could be turned into a distortion-free repeater by improving the purity of the vacuum.

Long-distance trials soon revealed the enhanced Audion tube to be the clear winner of four potential repeaters, and AT&T bought De Forest's patent in 1913. With the San Francisco Fair now delayed until 1915, they had two years to turn the trial into a working demonstration. This was no small effort, with four copper wires extending across the expanse of land; simply consider the work to erect the 130,000 poles that kept them up.

It was a colossal effort that also gave the astute marketing team at AT&T a unique opportunity to gain publicity. Alexander Graham Bell had founded the Bell Telephone Company in 1877, a year after his first and now legendary telephone call with his assistant Thomas Watson: "Mr Watson, come here – I want to see you." While the company no longer bore Bell's name, having merged with others to eventually become AT&T, he remained a familiar and popular figure in America.

For his 1876 telephone call, Watson had been in a different room from Bell but out of earshot. For this encore, Bell would be in San Francisco and his former assistant in New York. Echoing that conversation from almost four decades earlier, Bell said "Mr Watson, come here, I want you." To which Watson replied: "It would take me a week to get there now."

It was a triumph for AT&T on the national stage, but also a personal triumph for Jewett. His newly established research team, which now also included two more of Millikan's students, were breaking new ground for AT&T. And beyond. Millikan describes how in 1915, Arnold demonstrated wireless telephony across 200 miles. Later that year, they would broadcast from the Eiffel Tower in Paris to Honolulu.

In 1916 Jewett was appointed head of Western Electric, the manufacturing wing of AT&T. He commanded a team of roughly a thousand engineers, which included the research group headed up by Arnold. Jewett was relentless in his pursuit for new talent to join this group, writing to one of Millikan's former students – Harvey Fletcher – every year for five years before Fletcher finally caved in and joined the New York team. He was soon joined by Mervin Kelly, who had worked with Millikan on a crucial oil-drop experiment that both decisively proved the existence of electrons and attempted to measure their value.

The goal of the research group was to look beyond the day-to-day improvements and focus instead on the science. Under Arnold, it didn't merely tackle problems set out by senior management but performed research for its own sake – always with the communications in mind, but similar to research in universities the outcomes would not always be obvious. "Of its output, inventions are a valuable part," said Arnold,[5] "but invention is not to be scheduled nor coerced."

With business booming, so had the engineering department under Jewett, to the extent that in 1924 the AT&T board agreed that it should become its own company. Called Bell Telephone Laboratories, Inc., its staff of around 3600 employees would serve both AT&T and General Electric, with a budget of $12 million. That's around $220 million today, adjusted for inflation.

Jewett would later describe an industrial lab, of which Bell Labs was a prime example, as "merely an organisation of intelligent men, presumably of creative capacity, specially trained in a knowledge of the things and methods of science,

---

[5] Jon Gertner, *The Idea Factory*, p27

and provided with the facilities and wherewithal to study and develop the particular industry with which they are associated."[6]

This was in the midst of the Roaring Twenties, a decade that saw huge growth in the USA.[7] The country's gross national product increased by nearly 40%; the proportion of households with electricity shot up from 12% in 1916 to 63% by 1927; the number of cars on the road tripled. There was a feeling that the good times had come and would never end.

But, end it did, and dramatically so, with the stock market crash in 1929 that wiped a third off its value. The Great Depression quickly followed, with millions of Americans losing their jobs and subsequently cutting their phone subscriptions. AT&T's revenues plummeted, forcing Western Electric to lay off 80% of its workforce.[8] Bell Labs wasn't immune to this, with Jewett halting all hires, cutting pay, and instituting four-day working weeks.

Tragically, Arnold suffered a heart attack in early July 1933 at the age of 49.[9] His legacy lived on in numerous ways, however, not least through the Bell Labs Math department that he had instigated. This stemmed from his decision to hire a mathematician to support the engineers with their increasingly complex work. Arnold hired Thornton Fry for this in 1916, and it soon grew into a full department. One that George Stibitz joined in 1930.

By this time, the Math department had outgrown its support roots to being an active participant in research. Members were encouraged to use their brains inventively – and in Stibitz, they definitely had an inventor. In November 1937, as he explained in a 1967 article,[10] "I liberated some relays from a scrap pile at Bell Telephone Laboratories where I then worked, and took them home to start what I thought of as a play project."

He had become interested in the electromagnetic relays after being asked to investigate their magnetic behaviour. Whilst doing so, he was struck by the binary nature of the relay contact: "[it] had only stable conditions, off and on, and this I

---

[6] Jon Gertner, *The Idea Factory*, p32
[7] 'Roaring Twenties', *Encyclopedia Britannica*, **britannica.com/topic/Roaring-Twenties**
[8] Figures come from *The Idea Factory*, p36
[9] Harold DeForest Arnold', *Encyclopedia Britannica*, **britannica.com/biography/Harold-DeForest-Arnold**
[10] George R Stibitz as told to Mrs Evelyn Loveday, 'The relay computers at Bell Labs', in *Datamation*, April 1967, **rpimag.co/bellrelays**, p35

realised was true of the digits in binary notation".[11] This wasn't quite as obvious as might now be imagined. Even as a maths graduate, Stibitz explained, he had to look up the notation to familiarise himself with it because by the 1930s binary was seen as a historic curiosity.

Stibitz set to work at his kitchen table, fastening two of the stolen relays to a board. He then cut two strips from a tobacco can to serve as sheet metal, then nailed them to the board. These would be his input switches: press on either strip and the connection would become active. For output, he used flashlight bulbs. He then wired his machine so that if you pressed one sheet of metal, a bulb on the right would light up to indicate one; if you pressed both, it would add the two digits together and the left bulb lit up to indicate a carry. The other bulb would be off to indicate zero.[12] In time, this invention would be graced with its own name, the Model K Adder (the K short for kitchen), but its importance was by no means obvious at the time.

"I took it to the Bell Labs and showed it to some of my friends down there and they were satisfactorily amused by the idea that you could use binary notation from the old days to do arithmetic in the modern times, but then I wasn't entirely sure that it was so funny," he recalled in the lecture. What could he achieve with many such adding machines? Could it even mimic the mechanical desktop calculators that were then so prevalent at Bell Labs? He carried on the project during evenings at home, sketching out ever more elaborate circuits for multiplication and division.

All of this was a fascinating but theoretical exercise for Stibitz, but the invention that had so amused his colleagues coincided with a growing problem at Bell Labs: multiplying, dividing, adding, and subtracting complex numbers.

For those who have relegated all mathematics lessons to the farthermost reaches of their mind, a reminder about complex numbers. These include two parts: one is a real number, the other imaginary. By 'imaginary', we refer to the odd yet incredibly useful mathematical concept of the square root of -1. Which is, of course, impossible, as any numbers multiplied together will be positive. In maths, the square root of -1 is helpfully denoted by the letter 'i'. A complex number is therefore of the form 'a + bi', where 'a' is the real number and 'b' is the factor of 'i'.

---

[11] 'The Development, Design, and Use of the Bell Labs Relay Calculators, lecture by George Stibitz', Computer History Museum YouTube channel, recorded 8 May 1980, **youtu.be/FipGF3V9obU**

[12] You can see Stibitz demoing a replica in the YouTube video from around 11 minutes in, while the replica itself is now held by the Computer History Museum in Mountain View, northwest of San Jose, California.

Bell Laboratories Building, 463 West Street, New York City in 1925
Image: Charlesbahr, CC BY-SA 4.0

Complex numbers are particularly useful to express the characteristics of alternating currents, which play a crucial role in electrical engineering and telecommunications. Within the Bell Labs Maths department alone, Stibitz explained during the lecture, there were a dozen women who did "nothing but calculate with complex numbers using desk calculators."

The problem wasn't merely that this job was tedious and repetitive: calculations had to be written down to save them before being reused. These records, inevitably, were subject to human error[13]. Especially when you bear in mind that calculations needed to be accurate to several decimal places.

One member of Bell Labs proposed the solution of fastening together two calculators by mechanical means. "That plan was circulated around the laboratories and my then supervisor Dr [Thornton] Fry showed it to me and asked whether I had any ideas," said Stibitz. To which his answer was an emphatic yes: rather than mechanically attach calculators, you could connect them using electrical circuits, all based upon his relay-based Model K Adder. And this was how the idea for the Complex Number Calculator – as it came to be known – was born.

Stibitz set to work sketching out what the circuitry might look like. His schematic was handed to Samuel Williams, a relay circuit designer, in the hope that he

---

[13] The tedium behind long calculations, and the likelihood of mistakes, runs through this whole book, all the way from Babbage in our introduction to our final featured computer, the Pilot ACE. It also inspired the ENIAC, which was in effect 20 mechanical computers strung together and then automated.

would confirm that the idea could work. First, though, he had to decipher Stibitz's unorthodox notation and "peculiar squiggles".[14]

"He took the circuits that I had given him and actually traced through something like ten or twelve complex multiplications and divisions, step by step, through every one of the relay contacts and wires that I had drawn up," Stibitz told the Association for Computing Machinery in 1967, "and he came up with the conclusion that it would work."

Within a few weeks Bell Labs gave the go-ahead to the project, with Williams handed the job of creating the circuit designs while Stibitz kept charge of the theory. The bulk of time went into the planning, as the pair was breaking new ground. What kind of keyboard would it use? How much storage would it need? How many decimal places would it need to be accurate to?

All this planning took time, with the pair starting the work in September 1938 and only finishing it in April 1939. Remarkably, the manufacturing proved quicker than the planning. Under Williams's careful supervision, the computer was finished by October 1939. All they needed to do now was make it work.

This proved yet another big challenge. The Complex Number Calculator consisted of 450 telephone relays, filling roughly the same space as a tall American fridge-freezer; the larger left-hand side dealt with calculations for the real numbers while the right-hand side handled the imaginary part. "The two units operated in parallel," said Stibitz.[15] For example, during multiplication "the real and imaginary parts of the multiplicand were multiplied by digits of the real part and the multiplier simultaneously."

Problems occurred due to the intricate timing involved during calculations. If even one relay didn't activate quickly enough, the calculation would go wrong and nonsense would emerge. They ended up deliberately slowing down aspects of the calculation to ensure that the slower relays would complete in time, but the end result was still a fast machine by existing standards. Stibitz later recalled that addition would take around 0.1 seconds, while multiplication and division were likely to have taken around two seconds.

---

[14] 'Computer Oral History Collection, 1969-1973, 1977', part of the Association for Computing Machinery (ACM) Annual Meeting, 30 August 1967, **dl.acm.org/doi/10.5555/1234040.1234046**, tape 1

[15] George R Stibitz as told to Mrs Evelyn Loveday, 'The relay computers at Bell Labs', in *Datamation*, April 1967, **rpimag.co/bellrelays**, p39

The Complex Number Calculator was put into service in January 1940, at which point it could only do multiplication and division: these were the only functions that Stibitz and Williams thought were worthy of such a beast. However, user feedback quickly made it obvious that addition and subtraction would be heavily used, and by adding a few more relays – and some other relatively minor amendments – they were able to upgrade the machine. "This easy alteration, in so complicated a machine, pointed up its flexibility as contrasted with mechanical machines," said Stibitz.[16]

It was also easy to use, in part because operators didn't need to worry about binary notation. After all, if a PhD holder and maths graduate like Stibitz struggled to remember binary, it would be enormously off-putting to new users of his computer. While both Dr John Atanasoff with the ABC and Zuse with the Z3 invented a decimal to binary converter, so that users could think in normal numbers and the computers could work in binary, Stibitz took a different and innovative approach. In short, the Complex Number Calculator did both.

Its magic trick was to convert decimal numbers into a four-digit binary number. And it was actually even cleverer than this. In standard binary notation, you represent ten as 1010,[17] but this caused "considerable complication when we wanted to carry from one decimal column to another," said Stibitz.[18] To solve this, "I decided to shift all the digits by three units… Decimal zero is to be represented by what we would normally call a binary three, and so on up the line."

If zero is three, then that's 0011 (binary three), while ten would become its opposite: 1100. This symmetry became useful when handling negative numbers, but also meant that "a binary carry occurs at just the right time, namely when the sum reaches ten."

Not that the operators needed to worry about any of this. To use the computer, they sat at a desk with something resembling a typewriter in front of them. Rather than a normal keyboard, however, this included a row of numbers from 0 to 9 and operations in the row above. These allowed the operator to select the type of number

---

[16] George R Stibitz as told to Mrs Evelyn Loveday, 'The relay computers at Bell Labs', in *Datamation*, April 1967, **rpimag.co/bellrelays**, p43

[17] For those unfamiliar with binary, the 0s and 1s read from right to left and are multiples of factors of two. So, in 1010, it's 1 × 8, 0 × 4, 1 × 2, 0 × 1. Add those together and you have ten. 1111 would be 1 × 8, 1 × 4, 1 × 2, 1 × 1, which equals 15.

[18] 'The Development, Design, and Use of the Bell Labs Relay Calculators, lecture by George Stibitz', Computer History Museum YouTube channel, recorded 8 May 1980, **youtu.be/FipGF3V9obU**

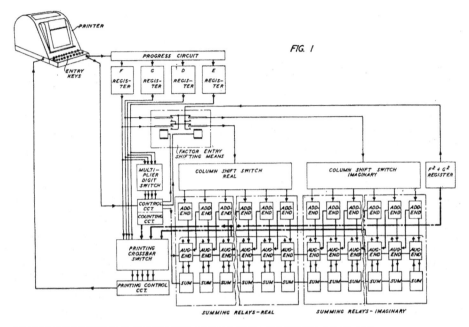

Schematic circuit diagram for the Complex Number Calculator from original US Patent filed in 1941 –
patents.google.com/patent/US2668661A/en
Public Domain

(real or imaginary) and to which operations they were applying: multiply, divide, add, or subtract. They could also call upon useful commands such as clear.

Stibitz's invention did have one quirk. Because it didn't support floating-point numbers, if an operator wanted to input 12.34, then she[19] needed to enter 0.1234 and make the decimal point adjustment after the result came through. For the skilled and experienced operators, however, this would have been trivial.

Once news of the Complex Number Calculator's brilliance spread, others within Bell Labs wanted access. Here, again, we see a sign of the company's innovative thinking: it installed two more input terminals on different floors. If the machine was engaged when someone wanted to use it, they would simply have to wait until it was free, just as they would if they called a telephone number and it was engaged.

---

[19] The operators were almost always women.

Fortunately, the computer took comparatively little time to calculate the results of even tough problems. "The speed of computation was phenomenal for that time: only 30 seconds or so for a complex division comprising three multiplications and addition, and two divisions in eight-place real numbers," remembered Stibitz in his 1967 article.[20] Once complete, the answer was printed out automatically on a teleprinter (often referred to as a teletype in the US due to the dominance of the Teletype Corporation, itself a subsidiary of Western Electric).

Bell Labs was never shy of publicity, and the Complex Number Calculator made a very public debut at the September 1940 meeting of the American Mathematical Society. This took place at Dartmouth College, Hanover, a good five-hour drive from New York City. And Stibitz had a great idea: rather than showcase its skills by telephoning problems through to an operator based at Bell Labs HQ, as Fry suggested, why not set up one of the teletype machines in Hanover?

One good reason is because it required AT&T to create a "28-wire teletype link between Hanover and New York, and coders and decoders were built at each end," according to computer historian Paul Ceruzzi in *Reckoners*.[21] Just in case anything went wrong, Sam Williams stayed behind in New York.

There was a lot at stake in terms of reputation, with the meeting on 11 September "attended by most of the prominent American mathematicians of the day," to quote Ceruzzi once more. He goes on to list John von Neumann and John Mauchly (co-creator of ENIAC) amongst the names.

First, Stibitz delivered a speech outlining why he had built the computer and what it was used for at Bell Labs. He then demonstrated the machine by running it through a few problems, later declaring that the "pauses during which the computer solved each problem were as impressive as the operating period."

All the problems were solved without any errors, but daringly Fry and Stibitz allowed all the attending mathematicians to try it for themselves. Norbert Wiener, an MIT professor who established the foundations for cybernetics (and coined the term),

---

[20] George R Stibitz as told to Mrs Evelyn Loveday, 'The relay computers at Bell Labs', in *Datamation*, April 1967, **rpimag.co/bellrelays**, p44
[21] Paul E Ceruzzi, *Reckoners: The Prehistory of the Digital Computer, from Relays to the Stored Program Concept, 1935-1945* (Greenwood Press, 1983, ISBN 978-0313233821), p92

tried to foil the computer by dividing by zero, but "was told by the machine that this was no good," said Stibitz.[22]

Viewed through the long lens of history, the Hanover demonstration was more than a simple technology showcase. A decade ahead of its time, it proved that distance was no barrier to users who wanted to access a distant computer's power, foreshadowing the practice of time-sharing – it would take another 24 years for General Electric to announce that all its 600 series computers would support time-sharing. Fittingly, it made that announcement at Dartmouth College.[23]

In its own historic account, Bell Labs even argued that after the demonstration "mathematicians from many parts of the country began, for the first time, to think seriously about new methods in computation."[24]

Journalists, who saw an early demo of the machine a few days earlier, were similarly impressed. "Computing Robot Solves Mathematical Problems Over Long Distance Wire" announced the headline in *The Birmingham Post* (Birmingham, Alabama, that is) on 10 September 1940.[25] "The answers rattled back on the machine in less time than any human being could solve them," wrote Watson Davis. "And they were free from the inaccuracies that human frailty sometimes commits."

He continued: "For the present at least you will not be able to dial 110 and ask long distance to do your math problems for you. The electrical computer was made for use on the Bell Laboratories' own problems. The one machine constructed so far, when it gets through its demonstrations, will be fully occupied with real computing work already in sight."

Those words were prescient. Aside from the occasional day off for maintenance, the reliability of relay switches and the thorough debugging done by Stibitz and Williams before the machine went live meant it put in long hours of service, reliably, until it was finally decommissioned in 1949. And Davis was also correct in hinting that more machines were to come, even though that would not have been obvious in 1940.

---

[22] 'The Development, Design, and Use of the Bell Labs Relay Calculators, lecture by George Stibitz', Computer History Museum YouTube channel, recorded 8 May 1980, **youtu.be/FipGF3V9obU**

[23] James Pelkey and Andy Russell, 'Timesharing – Project MAC - 1962-1968', The History of Computer Communications, **rpimag.co/timesharing**

[24] Bernard D Holbrook and W Stanley Brown, 'A History of Computing Research at Bell Laboratories (1937-1975)', 1982, **rpimag.co/bellhistory**, p6

[25] Watson Davis, 'Computing Robot Solves Mathematical Problems Over Long Distance Wire', in *The Birmingham Post*, Alabama, 10 September 1940, p3

For, despite its success, the executives at Bell Labs felt the computer was simply too expensive. "The Complex Calculator was then studied by the group at Bell Laboratories and everyone was horrified to find that it had cost $20,000 [$450,000 in today's money] including the development, design, construction, and debugging," said Stibitz.[26] "The idea that anybody should spend $20,000 for a mere calculator was something that was not acceptable, and the Labs decided that no more things of this sort would be built."

Despite this, Stibitz continued to think about improvements for further machines. One idea was to introduce support for floating decimal points. Another was to add a self-checking code to isolate errors with particular errors; as it was, the mechanics used toothpicks to isolate errors during debugging checks. Similarly to the Harvard Mark I – which Stibitz knew nothing about due to the high levels of secrecy – he wanted to add a tape input so that they could vary the type of programs the computer ran.

These ideas would likely have come to nothing had it not been for the outbreak of war, with the US finally joining the rest of the Allies in December 1941 after the Japanese attacks on Pearl Harbor. Stibitz was loaned out by Bell Labs to the National Defense Research Committee (NDRC), and would go on to create four wartime computers based on relays.

The first, initially called the Relay Interpolator but later renamed the Model II, tied in with an analogue computer also designed by a Bell Labs engineer. The M9 gun director worked in tandem with a radar system to automatically track aircraft, but to work effectively it needed to be trained. This required a huge number of calculations, and rather than rely on desk calculators Stibitz suggested they build a relay machine to do the job.

This time Ernest Andrews "took over the technical work of designing it and having it built," said Stibitz.[27] "This machine had a tape program and it would spend most of its working hours, nights and Sundays, loaded up with enough tape to punch output data and enough paper to produce a wastebasket full of information which we were using in the War program."

---

[26] 'The Development, Design, and Use of the Bell Labs Relay Calculators, lecture by George Stibitz', Computer History Museum YouTube channel, recorded 8 May 1980, **youtu.be/FipGF3V9obU**

[27] 'Computer Oral History Collection, 1969-1973, 1977', part of the Association for Computing Machinery (ACM) Annual Meeting, 30 August 1967, **dl.acm.org/doi/10.5555/1234040.1234046**, tape 1

It included a similar number of relays to the Complex Number Calculator, later renamed the Model I in light of all its successors, and could only perform addition and subtraction. Technically it could perform multiplication by a small integer, but it used repeated addition and was slow, taking four seconds. Still, it was the first of Stibitz's relay machines to be programmable by tape and due to that flexibility it stayed in service from July 1943 until it was dismantled in 1961.

According to the official Bell Labs history,[28] the people of Britain have much to thank these anti-aircraft guns for: "During the month of August 1944, over 90 percent of the [V-1] buzz bombs aimed at London were shot down over the cliffs of Dover."

The Ballistic Computer, Model III, was more ambitious still. Completed in June 1944, this packed in 1400 relays and was capable of far more powerful calculations than its predecessor. Its main function was again to train anti-aircraft gun directors, with Bell Labs' Joseph Juley reassuringly stating that "shells obviously cannot be fired at our own planes" in his detailed article about the computer.[29]

Instead, planes would go on trial runs and be tracked by the gun director, with measurements such as elevation tracked every second. "Since such a trial run may last as long as 200 seconds, and for each second a large number of computations are required, it would take a team of five [human] computers at least a week to obtain the results desired," Juley added. The Ballistic Computer could perform the same calculations in five or six hours.

The Model III was completed in June 1944, and a year later was followed by the Model IV – or, clumsily, the Error Detector Mark 22 – in March 1945. With almost identical specifications to the Model III, computer historian Paul Ceruzzi argued that they had "just the right balance of features for their relay technology." And both would do sterling service for their military masters, with the Model III retiring in 1958 and the Model IV in 1961.

But arguably the most interesting of all the Bell Labs relay computers is the Model V. The company made two copies, one for the US Army and one for

---

[28] Bernard D Holbrook and W Stanley Brown, 'A History of Computing Research at Bell Laboratories (1937-1975)', 1982, **rpimag.co/bellhistory**, p3
[29] Joseph Juley, 'The Ballistic Computer', in Brian Randell (ed.), *Origins of Digital Computers: Selected Papers* (Springer Verlag, 1982, ISBN 978-3540113195), p257

aeronautical research.[30] They weren't completed until after the war, by which time Stibitz had left the NRDC, but it was in many ways the pinnacle of his vision. It was also huge and expensive, with more than 9000 relays and an estimated cost of $500,000. That's well over $8 million in today's money.

You could argue that each Model V was two computers rather than one, though, as they featured two separate arithmetic units that called upon their own memory registers and input-output devices. So operators could run two smaller problems independently or take advantage of both processors (to hijack the more modern term). At a push, you could even say it had its own operating system, with a control unit that dictated the workflow to each processor. Stibitz called this "superbranching", so the computer even wins for hyperbole.

There would be one last hurrah for Stibitz's relay computers in the Model VI, which Bell Labs built for its own use in 1949. This was a stripped-down version of the Model V, with only one 'processor', but it looked flat-footed compared to the electronic rivals sprouting up around it. Ceruzzi describes it as "at least an order of magnitude below even the slowest electronic computers", and even with the inclusion of conditional branching it simply couldn't compete with later rivals.

Before we leave Bell Labs, though, we should give due tribute to other contributions to early computing. One is the prioritisation of reliability, something very much lacking in other computers in this book. The longevity of all the relay computers is testament to this, but later models also included error detection – and even error correction.

The problem stems from intermittent errors, such as a dust particle getting in the way of two relay contacts. So, for a few cycles, the calculations will simply be wrong. Then the dust is dislodged and the relay works again. This is quite unlike valves (vacuum tubes): when they failed, they failed for good. That means that even though Bell Labs established a diagnostic procedure to periodically check for faults, it might miss an intermittent problem and not know that it should discard results.

That's why Bell Labs quickly designed independent circuits that would check calculations at "every stop of a computation, like policemen looking over everyone's

---

[30] The army model went to Aberdeen Proving Ground, as featured in the story of the ENIAC, while the other model was used by the National Advisory Committee on Aeronautics' laboratory at Langley Field, Virginia.

shoulder," wrote Ceruzzi.[31] From the Model II onwards, they also used seven binary relays rather than four so that they could build in checks. We've tried to avoid printing cumbersome tables in this book, but here we can't resist:

Decimal digit	'Weight of five'	Single relay positions
0	01	00001
1	01	00010
2	01	00100
3	01	01000
4	01	10000
5	10	00001
6	10	00010
7	10	00100
8	10	01000
9	10	10000

So, it's binary but not as we know it. Instead, each decimal digit is represented by two active relays to indicate its multiple of five and how many ones it has. What's clever is that for any number, only two relays will be on: one in the first group, one in the second group. If three are on, or one, the testing circuit knows there's a problem. A separate checking circuit made sure that only one relay was on in each group.

From this clever acorn came the mighty oak tree of parity checking, which was devised by Bell Labs' Richard Hamming and works in a similar way. As in, it checks that the right number of binary elements are active. "Hamming went on to show that if more redundancy was added to a number code, it could not only detect errors but

---

[31] Paul E Ceruzzi, *Reckoners*, p97

Relay equipment room of the Model V – a successor to the Complex Number Calculator – installed at Ballistics Research Laboratories, Aberdeen, Maryland
US Army photo, Public Domain

also *correct* them as well," wrote Ceruzzi. "[His] work has formed the basis for nearly all computer circuit design ever since."

Our final mention goes to a potentially familiar name in William Shockley, who joined Bell Labs in the mid-1930s after gaining his PhD from MIT. Shockley started his training in the vacuum tube laboratory, he explained in a 1969 interview,[32] "and I was given a lecture by the then research director, Dr Kelly, saying he looked forward to the time when we could get all the relays that make contacts in the telephone exchange out of the telephone exchange and replace them with something electronic so they'd have less trouble."

Shockley had earned his doctorate due to his research in solid-state physics, and Kelly's remarks inspired him to work out a way to use solid-state technology to create a rival to relays. By December 1939, he had developed what he considered a working

---

[32] William Shockley interview by Jane Morgan, Palo Alto Historical Association, 1969, **youtu.be/LWGVuoisDbI**

56 The Computers That Made The World

transistor, only for war to intervene. On his return he picked up where he left off, building on insights about crystals (as the semiconductor materials) he gained during the war to help.

History records the birth of the transistor as 1947, with the shared credit (and Nobel Prize in Physics) going to Shockley and his Bell Labs colleagues John Bardeen and Walter Brattain. While Shockley's name has gone down in computing history[33] in a way that George Stibitz's name has not, we hope that, like a dust mite interrupting a relay connection, this is merely an intermittent fault.

---

[33] Albeit with considerable controversy over his extremist views on race and eugenics.

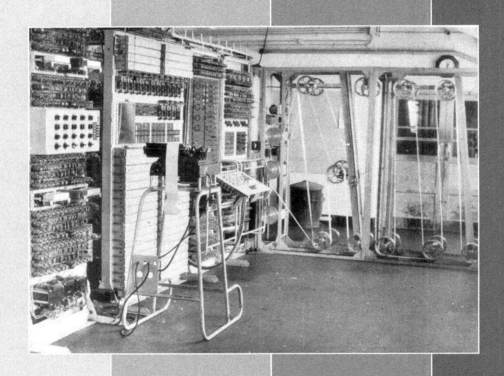

The tenth Colossus computer with its extended bedstead in Block H at Bletchley Park in 1945

Image: UK Public Record Office, Public Domain

Colossus

"Like magic and science combined!"
Joanna Chorley, Wren

Perhaps it is poetic justice that the world's first large-scale electronic digital computer was shrouded in secrecy. The Colossi of ancient times – the Greek word κολοσσός (kolossós) literally means 'gigantic statue' – are the stuff of myth and legend, whether that's the lost Colossus of Nero or the Colossus of Rhodes destroyed by an earthquake in 226 BCE.

The Colossus in this story fell foul of British national secrecy rather than natural elements. With everyone involved sworn to silence under the Official Secrets Act – revealing the computer's existence would have been an act of treason – its key role in breaking Nazi High Command codes was kept under wraps until the 21st century.

Even today, few people understand exactly what the Colossus did, how it worked, or the role played by the men and women behind the machine. Perhaps Tommy Flowers has earned name recognition – and rightly so – but there are dozens of others who were key to its success, whether that's code breakers such as Bill Tutte and Max Newman, or electrical engineers such as Sid Broadhurst and Bill Chandler.

Then there are the hundreds of unsung heroes whose names are rarely recorded in history books. The dozens of people who worked under Tommy Flowers, executing his plans. The factory workers who built the Colossus – more accurately, the Colossi, as ten were in operation by the end of war. The hundreds of Wrens (the women's branch of the Royal Navy) and their army counterparts, or the radio signal interceptors who painstakingly recorded the German transmissions. As they had to: one mistake could render the whole message useless.

There is also one giant hero, which is Britain's wartime General Post Office. It was the GPO rather than Bletchley Park that took the risk to build the first Colossus, and this insufficiently honoured organisation built the infrastructure on which all wartime communications relied.

But let's start the story of the Colossus on 1 September 1939. Tommy Flowers was then a 33-year-old employee of the Post Office's research department and had just arrived in Berlin, of all places, to take part in a European conference about telephone systems. Germany invaded Poland the same day.

Flowers and a fellow British delegate went to the British embassy to let them know where they were staying. "The young man at the embassy ... looked at us and said, 'You people must be mad, there's going to be a war in a few days'," Flowers

A Colossus Mk 2 being operated by Dorothy Du Boisson and Elsie Booker, 1943. The slanted control panel is on the left, with the 'bedstead' paper tape transport on the right
Image: The National Archives (UK), Public Domain

said in a far-reaching interview in 1998.[1] "Which surprised us a bit. But we couldn't go back."

The British duo spent the next day with their German counterparts on the committee, trying to play it cool, but the following morning the embassy called to say they had to catch the next train out or else. Flowers said of the tense eight-hour journey through Germany: "We hardly saw anything, there was hardly anyone on the stations, but when we finally got to Holland, which was after dark, everywhere was in a blaze of light. The Dutch army was mobilising, the same in Belgium."

After catching the boat train across to England, they arrived at Liverpool Street station at 8am on 3 September 1939. Flowers sensibly exchanged his ten days' worth of German Marks into Pounds at the first opportunity, while his fellow traveller decided to wait until 11am. At which point the UK had severed diplomatic ties with

---

[1] Thomas Harold Flowers interview with Peter Hart, 18 May 1998, Imperial War Museum Collections, **iwm.org.uk/collections/item/object/80017376**. All quotes from Tommy Flowers are taken from this interview unless otherwise stated.

Germany and the banks stopped taking its currency. "He never got his money back," said Flowers.

Still, it had been a lucky escape for both men. Many British citizens unfortunate enough to be in Germany on the day Britain declared war were detained at internment camps until VE Day in May 1945. It was an even luckier escape for the British war effort, because without Tommy Flowers there would have been no Colossus – and without Colossus the Allies' path to Germany would have cost many more lives, on both sides.

To understand why Tommy Flowers was so intrinsic to Colossus, we now need to go back to 1922 when he left school at the age of 16. He had been a star pupil and "was determined to be an engineer", just like two of his uncles; their advice was to "go through the workshops" rather than into further study. So he took up an apprenticeship at the Royal Arsenal in Woolwich, but also enrolled in engineering evening courses at Woolwich Polytechnic[2] (now the University of Greenwich).

In 1926, by which point Flowers was still only 20, the General Post Office[3] decided to accelerate its move to automatic telephone exchanges. While automatic exchanges were by no means a new concept, and had been rolled out in several British cities, many calls still relied on physical connections being made by operators sitting at a switchboard. When a call came through, they had to manually connect callers with cords.

Desperate to modernise, the Post Office needed engineers and they needed them fast. So they invited school leavers to take an examination, with the best performers invited to join the scheme. And Tommy Flowers, it turns out, was the best of the lot, beating hundreds of others to top the exam scores.

Within three months of starting, Flowers said he was "yanked out of" general training and sent to the GPO's Circuit Laboratory. It was Flowers's job – along with everyone else in the lab – to design new circuits based on electromagnetic relays that could be rolled out in exchanges. Much of his work was akin to programming today, but rather than coding in software Flowers was coding in hardware. First it was a

---

[2] See 'Thomas "Tommy" Flowers MBE, 1905-1998', University of Greenwich alumni, gre.ac.uk/portraits/alumni/thomas-tommy-flowers-mbe-1905-1998
[3] The General Post Office took ownership of the National Telephone Company in 1912, because the ruling Liberal government had decided the NTC had become a monopoly (having bought up the local, private telephone companies in the 30 years between 1881 and 1911).

matter of designing the circuits, then testing them to see if they worked as desired. Again, a familiar scenario for today's programmers, and excellent training when it came time to build the Colossus.

The Post Office installed London's first automatic exchange in Holborn in 1927, and by the time Flowers left the Circuit Laboratory in 1930 "we were well advanced, we got a lot of exchanges working, and they were working reasonably well." He left to join the Post Office Research Station in Dollis Hill, where he was assigned to work with a senior engineer who was studying the problem of long-distance dialling.

"With ordinary electricity in wires, the limit to the distance is about 40 miles," said Flowers. Voices would become too faint to hear. "This chap I was put with was saddled with the problem [but] wasn't getting anywhere because he wasn't any good … but after about six months, during which time I almost went barmy, he was then promoted. If you had someone on your staff who wasn't any good, the only way to get rid of him was to get him promoted."

When Flowers was given the job to solve, he turned to thermionic valves. He realised that not only could these replace electromechanical switches, and work thousands of times faster, but they could be used to amplify people's voices. This wasn't a unique insight – unbeknownst to Flowers, Bell Labs had almost exactly the same idea in the USA, as we detail in Chapter 3 – but he was the first person to introduce valves for this purpose in Britain. The trials took place between London and Bristol, roughly 120 miles, and Flowers remembered that one of the first 'tests' was for him to call his then fiancée and future wife, Eileen Green.

By the outbreak of war in 1939, his valve-based system had moved out of the lab and into production. The Research Station also had a new boss in the form of Gordon Radley, who Flowers described as a "warped genius". Radley had an astounding memory for detail and he needed it, as the Dollis Hill team stopped all civil tasks and focused instead on the war effort.

Flowers spent the early part of the war working out the best way to transmit data from radar stations on the coast to Fighter Command in Stanmore, Greater London. It wasn't until early 1942 that he signed the Official Secrets Act and was informed of the code-breaking work being done at Bletchley Park. From that point on, revealing anything about his work with the Government Code and Cipher School (GC&CS) would be an act of treason.

The reconstructed Colossus at The National Museum of Computing, Bletchley Park
Image: Ted Coles, CC BY-SA 4.0

Soon after putting pen to paper, an act that effectively stopped him even talking about Colossus for decades, Flowers met Alan Turing for the first time. "[Turing] explained the technology of code breaking," said Flowers. "They needed the equivalent of the [Enigma] coding machine … they wanted something that the girls could operate with a sort of typewriter keyboard."

Flowers went away and designed a machine, but it was much bigger than the Bletchley Park team was expecting – based on large electromagnetic switches, as that provided reliability – and meanwhile the German Enigma machines had evolved. "When they got it they were disappointed, they didn't want it," said Flowers. Despite this, Turing was impressed by Flowers's engineering skills, and Radley's wartime diary[4] reveals that Flowers would sometimes accompany his boss to Bletchley Park from that point on.

We should briefly pause to mention a side of Britain's code-breaking weaponry that rarely gets a mention: the radio listening stations, called Y stations, that intercepted enemy communications. "They were kind of the front wall, they were picking up the messages, and if it hadn't been for them, there would have been nothing for the code breakers to break," says Jack Copeland, Professor of Logic at the University of Canterbury in New Zealand, and a world authority on the Colossus.

One of those visits concerned a different code-breaking challenge: a new type of signal had been picked up by one of the Y stations in the latter half of 1940,

---

[4] The diary is available for download from **bt.com/about/bt/our-history/bt-archives/dollis-hill**

with more following in early 1941. Quite unlike Enigma, which transmitted in Morse code, messages of this new type were sent as teleprinter code. Here, each character is represented by a five-strong pattern of pulses and non-pulses. Y, for instance, is pulse-gap-pulse-gap-pulse. Or 10101 in binary notation. At Bletchley Park, they denoted pulses as crosses and gaps as dots, so Y becomes x • x • x.

Sophisticated automatic teleprinters capable of reading and transmitting at high speeds had been around since the start of the 20th century, so when these radio messages started appearing in 1940, people had no problem recognising their distinctive tones. Initially, however, their existence was merely noted. There was too much else to do.

It was only in June 1941, when an RAF station near Folkestone intercepted an unencrypted message revealing military details, that Britain's network of intercept stations began to record them. Naturally, these messages – now encrypted with an unfamiliar cipher – were sent to Bletchley Park for examination.

While Bletchley Park had no idea of the cipher being used at this point, nor even the machinery, analysis of radio signals provided useful information. "Some of the bigger Y stations carried out radio fingerprinting," says John Pether, one of the volunteers who looks after the Colossus replica at The National Museum of Computing, adjacent to Bletchley Park. "They used a specialist oscilloscope and passed a 35mm film over the screen at high speed. They could determine whether the transmitter was running off batteries, what the frequency domains were, whether the key contacts were dirty to identify individual keys."

But the real prize was to decipher the messages' contents. Within two months, this rising challenge was being met by Bletchley's Research Section, reporting to the already legendary Colonel John Tiltman. Tiltman found his way into code breaking through the army. A World War I veteran, his ability to pick up languages became clear when he learned Russian without formal training whilst serving in a British contingent with Belarusian soldiers. Originally seconded to work for the newly created Government Code and Cipher School as a Russian translator in 1919, the move was rapidly made permanent when his incredible ability to spot patterns became apparent.

Alongside Dilly Knox, another famed cryptologist who would play a crucial role in breaking the Enigma ciphers, Tiltman broke Russian Comintern signals[5] in the early

---

[5] Lenin established the Comintern, short for Communist International, in 1919. Its aim: to promote global communist revolution and unite communist parties under a single banner.

1920s. In 1933 he turned his attention to Japanese military codes, breaking six of their cipher-systems, so by the time he was put in charge of the Research Section his legend was already made. As was his reputation as one of the nicest men at Bletchley.

"My arrival was unforgettable," recalled William Filby of his first encounter with Tiltman.[6] "As I saluted, I stamped the wooden floor in my Army boots and came to attention with another shattering noise. Tiltman turned, looked at my feet, and exclaimed: 'I say old boy. Must you wear those damned boots?'" From that point on, added Filby, he would wear his full army uniform with a pair of white running shoes, much to the exasperation of more formal army officers.

An Enigma I machine in the Museo Nazionale Scienza e Tecnologia Leonardo da Vinci, Milan
Image: Alessandro Nassiri, CC BY-SA 4.0

By the end of August 1941, through Tiltman's work, Bletchley Park knew several things about these non-Morse broadcasts. Crucially, they knew they were encrypted using the Vernam cipher, which added a "random" value to each of the five dots or crosses (called impulses) that made up a character in the teleprinter code. As a reminder, Y is represented by five impulses: x • x • x. If the encrypting key was J, represented by x x • x •, then the new enciphered character would be the sum of its parts.

To understand this process, it may be easier to use binary notation. Here, Y is equivalent to 10101 (x • x • x) and in our example the encrypting key is J, or 11010 (x x • x •). Adding the first impulses together means adding x to x. Or 1 to 1. The answer

---

[6] This quote is taken from Michael Smith's excellent book, *Station X: The Codebreakers of Bletchley Park* (Pan Books, 2004, ISBN 978-0330419291), p84

is •, or 0 (note there is no carryover to worry about here, which is why 1 plus 1 is 0). For the second component, we're adding x to •. That sum is x or 1. Going through all five components, Y + J is 10101 plus 11010. So the broadcast letter is represented by 01111, otherwise known as the pulses • x x x x . That is, the letter T.

One more crucial thing to understand about the Vernam cipher the Nazis were using is that you decrypt by applying the same key. So, adding T to the encrypting key J (breaking this down to impulses, adding • x x x x to x x • x •) gives you Y (x • x • x).

While at this point Bletchley Park had no idea what the encrypted messages said, they knew they were being sent from Berlin to a German military outpost in Greece. This alone made them interesting. They called the radio link Tunny, then a common name for tuna, and would continue the fish-based code names as other connections sprang up. For example, the Berlin to Rome connection was called Bream, distinguishing it from Berlin to Copenhagen (Turbot) or Rome to Tunis (Herring).

"Each message was preceded by a string of twelve letters, which Bletchley Park called the indicator, which functioned to tell the recipient of the message the positions to which they should turn the wheels of their machine in order to decrypt the message," explains Copeland.[7]

However, this information was initially of little use to the code breakers as they had no idea of what ciphering machinery was being used to generate the encrypting key. How was it wired? How did the wheels turn? Were there any weaknesses the Bletchley Park team could exploit? They needed a piece of luck.

That came on 30 August 1941 when the same twelve-letter indicator, HQIBPEXEZMUG, appeared at the beginning of two separate messages. Two or more messages with the same indicator were called 'depths' and, as we shall see later in this chapter, depths played a crucial role in Bletchley Park's attack on Tunny messages.

If the operator had simply resent the same message, it would have been no help. Here, however, the messages started the same but soon differed, with the second message around 100 characters shorter in total. This suggested it was intended as a resend of the first message, but that the operator had begun abbreviating words to save himself time. Which it no doubt did: but it also meant Bletchley Park could add the messages together and eliminate the encrypting key.

---

[7] Interview with author, as are all direct quotes in this chapter from Jack Copeland unless stated.

As explained above, this key effectively cancels itself out when two characters are added together.

This meant the British code breakers had a stream of deciphered letters that were the result of two combined messages. Now all they had to do was untangle them. Tiltman set to work, and after several days had not only revealed the plaintext[8] of the two messages, but also the whole encrypting key – almost 4000 characters long – that could surely help them work out the structure of the machine.[9]

Were that it was so easy. Three frustrating months followed in which the Research Section failed to make progress. Enter Bill Tutte, who shared his part in the story in *Colossus: The Secrets of Bletchley Park's Codebreaking Computers*, edited by Jack Copeland. "Was it a gesture of despair that Captain [Gerry] Morgan, that day in October, handed me the Tunny key, with associated documents, and said 'See what you can do with this?'"[10]

Tutte, it turned out, could do a lot. He started with the hypothesis that twelve indicators meant twelve wheels. And since eleven of the twelve indicators used all letters in the German alphabet (aside from J), but one only used 23 letters, might one of the wheels have 23 settings? We won't go into detail here, for it's enough to know that this supposition was right and that, combined with many other suppositions – and a huge amount of trial and error – Tutte discovered patterns. And if there's one thing cryptographers love, it's patterns.

With the help of other members of the Research Section, who joined in once Tutte's method of attack bore fruit, they brilliantly deduced the inner workings of the machine. From those 4000 characters of encrypting key, they worked out that the wheels came in two sets: what they called psi-wheels and chi-wheels (mathematicians commonly use Greek letters). Both sets (psi and chi) included five wheels, with two motor wheels in the centre.

They also knew that the wheels featured movable 'cams'; that is, mechanical components that could be slid by hand from an active to an inactive orientation by

---

[8] That is, unenciphered text. We'll use the term plaintext to distinguish it from enciphered text.
[9] Tiltman's work is set out in detail in 'The Tiltman Break' by Friedrich L Bauer, in Jack Copeland (ed.), *Colossus: The Secrets of Bletchley Park's Codebreaking Computers* (Oxford University Press, 2006, ISBN 978-0192840554), Appendix 5, p370
[10] William T Tutte, 'My Work at Bletchley Park', in Jack Copeland (ed.), *Colossus: The Secrets of Bletchley Park's Codebreaking Computers*, Appendix 4, p356

pushing them sideways. More than that, they knew exactly how many cams each wheel contained. And that while the starting positions for each wheel were changed with every message, the cam orientations would stay fixed for weeks on end.

At Bletchley Park, they distinguished between the starting position of the wheel and the orientations of the cams on the wheel by describing the first as the wheel *settings* and the second as wheel *patterns*. Once the wheel patterns were changed, the team was locked out until a German operator went against protocol and sent two messages using the same indicator. That is, a depth.

It's here, in July 1941, that Alan Turing makes his solitary but pivotal appearance in the story of Colossus. In Tutte's simple words, "Turing became interested in the problem of breaking a true Tunny depth and he found a method of doing so."[11]

The key to Turingery, as the method became known, was to examine differences determined by the number of cams on a wheel. For example, Turing knew that the first chi-wheel included 41 cams and so whatever position each cam was in – on or off – would repeat every 41 characters. There would be a pattern, albeit a deeply hidden one.

Fortunately for the code breakers, the German design added a 'stagger' where a psi-wheel might stay in the same position. The idea was to add complexity, but it actually introduced a weakness. By examining the differences between encoded characters, once broken into a series of dots and crosses, Turing showed that it was possible – through a mix of luck, a giant sheet of graph paper, much crossing out, and sheer repetition – to uncover each cam's orientation on one wheel, and then to determine the orientations of all the cams on all the wheels.

Turingery was a laborious process, but, at this stage of the war, the wheel patterns were only changed once a month or longer (to be accurate, the chi-wheels were changed monthly, the psi-wheels every three months).

"Once Turing had shown the Research Section how to extract the wheel patterns, then they were in the game," says Copeland. "They'd got exactly the same information the German guy who was decrypting the message with his Tunny machine had got. They knew the wheel patterns and the message indicator was telling them what the

---

[11] William T Tutte, 'My Work at Bletchley Park', in Jack Copeland (ed.), *Colossus: The Secrets of Bletchley Park's Codebreaking Computers*, Appendix 4, p360. Jack Copeland gives an overview of how Turingery works from pp380-382.

wheel settings were, so they could simply set up their replica machine and decrypt the traffic, just like the German recipient."

With a procedure laid out, the day-to-day job of breaking Tunny messages moved to a section run by Major Ralph Tester. The Testery, as it was called, could now decipher a torrent of messages encrypted using Tunny until the wheel patterns changed, so long as the Nazis were kind enough to keep sending the indicators at the start of each message.

This generosity continued until November 1942, when the German military switched instead to numbers found in a book issued to Tunny operators. These numbers told the operators the starting positions for the wheels. At which point, the Testery was plunged into near-darkness as the torrent became a trickle. All it could hope for was a depth (two messages sent with the same indicator), which would allow them to apply Turingery, but this time 'in reverse' to extract the wheel settings.

Crucially, however, the Nazis had tightened up other security measures too. "It was a double whammy," says Copeland. "The code breakers lost the twelve-letter indicators, and the depths were getting rarer, so the number of messages they could break using Turing's method was becoming smaller and smaller."

Bill Tutte was not to be deterred. Building upon the same ideas as Turingery, but introducing statistical methods, he devised a way to work out the starting positions of the first and second chi-wheels for any message. His so-called '1+2 break' meant that with enough time the code breakers could detect the most likely chi-wheel start positions simply by counting (and adding) dots and crosses.

The idea was to add the teleprinter message to the stream of chi (that is, every possible combination of encrypting key from the chi-wheels), then count up the dots. Then slide it along by one and repeat the process. And so on until you had tried all the different possible chi streams. Tutte worked out that if around 54% of the individual pulses were zeroes when totalled up, the wheel settings were almost certainly correct.[12]

---

[12] Interested readers should read in-depth descriptions found in the technical appendices of *Colossus: The Secrets of Bletchley Park's Codebreaking Computers*. But here's a quick summary. In teleprinter code, as we detailed above, each letter can be represented by a five-long string of dots and crosses, or zeroes and ones. Each letter in the encrypting key is also represented by five dots and crosses. These are added together to form a new character, the ciphertext, which is what is transmitted. Tutte's breakthrough was to realise that the correct settings for the first pair of wheels would be revealed (probably but not every time) by a very slightly higher number of dots – or zeroes – in the resulting stream of five-digit code. This is why it became a counting game.

But the sharper-eyed reader would have spotted the phrase 'enough time' above, which hides a large problem. 'Enough' here, according to Newman's own estimate, could mean centuries for some messages. Even though they only needed to deduce a pair of wheel settings to begin the process, that meant 1271 (41×31) different scans of the text to spot likely candidates. And as this was a statistical method, they needed long messages, not short ones, adding to the time.

All of which meant they needed a much quicker way to count and add – and that meant some form of automatic adding machine. Especially as, even after the first two chi-wheel positions were known, you would have to repeat the method to reveal the three remaining chi-wheel settings. Only at this point would experienced code breakers have enough information to work out the remaining wheel positions by hand.

Tutte took his 1+2 break theory to Captain Gerry Morgan and Max Newman. "They began to tell me, enthusiastically, about the current state of their own investigations," wrote Tutte.[13] "When I had an opportunity to speak I said, rather brashly, 'Now my method is much simpler.' They demanded a description. I must say they were rapidly converted."

In what historian Jack Copeland describes as "a moment of inspiration",[14] Newman realised that one of his Cambridge colleagues, Eryl Wynn-Williams, had already designed an electronic counter based on valves that he thought could be adapted for this crucial part of the job. He called in Wynn-Williams to take control of that side of the design, with the GPO in charge of creating the drive mechanism for the tapes, the photoelectric readers to detect dots and crosses as they passed, and a combining unit to add up the numbers.

But when we say the GPO was put in charge, we should emphasise that it wasn't given to Tommy Flowers: instead it passed to Francis Morrell, head of the teleprinter group. Morrell commissioned GPO engineers Eric Speight and Arnold Lynch to produce the photoelectric readers (without telling them why).[15] They did an excellent job, but the combining unit – which used electromagnetic relays – proved more challenging for Morrell and his team. It was time to call in an expert in relays, Tommy Flowers.

---

[13] William T Tutte, 'My Work at Bletchley Park', in Jack Copeland (ed.), *Colossus: The Secrets of Bletchley Park's Codebreaking Computers*, Appendix 4, p364

[14] Jack Copeland, 'Machine against Machine', in Jack Copeland (ed.), *Colossus: The Secrets of Bletchley Park's Codebreaking Computers*, Chapter 5, page 64

[15] Brian Randell, 'Of Men and Machines', in Jack Copeland (ed.), *Colossus: The Secrets of Bletchley Park's Codebreaking Computers*, Chapter 11, p146

Valves (vacuum tubes) in the reconstructed Colossus in the National Museum of Computing, Bletchley Park
Image: CGP Grey, CC BY 2.0

"I quickly came to the conclusion that the [machine] would never be any good, that we were wasting time on it," said Flowers. In particular, he was extremely sceptical about keeping two tapes in perfect synchronisation. One run with a long message, he pointed out, could take several hours and that would surely be too much for paper tape to stand. "So I thought up something different."

Flowers's key idea, one he was uniquely placed to come up with, was that by using thermionic valves as switches "an analogue of the coding machine could be made". So no need for tape, as the data would be stored within the machine.

While Flowers was convinced his solution would work, he struggled to convince the Bletchley Park hierarchy. This was partly due to thermionic valves' poor reputation for reliability – used in radio sets and early TVs, they were vulnerable if moved and were well known for their limited operational life – but Flowers had built automatic exchanges with hundreds of valves in them, and knew that they would prove extremely reliable so long as they were kept switched on.

Newman appears to have been convinced on this point, but there was a second problem. When Newman asked Flowers how long it would take to build, the Post Office engineer gave the honest reply of at least a year. "[Newman] said, well, they just couldn't wait a year; in a year the war could be over and lost," said Flowers.

Despite this, Newman cautiously welcomed Flowers's ideas. In a report to Commander Edward Travis, dated 1 March 1943, he wrote:[16] "Flowers, of the PO, has produced a suggestion for an entirely different machine, in which the message, and the wheels to be compared with it, would be set up on valves, by means of relays. This would involve 5000 or so valves, and about 2500 relays. (The heat the valves would generate would be equal to that of eight two-bar electric fires!)"

While Newman describes this as an "ambitious scheme", he isn't against it. "I feel that this is basically the right sort of approach, and that it is very much to our advantage to try out these techniques, and if possible get a step ahead with them. At the same time, since there is a risk of hold-ups along these new paths, I emphasised that the simpler tape-machine (which also has the advantage of easy adaptability to all sorts of purposes) should be gone on with also, at full speed."

One of Newman's criticisms of Flowers's proposal was that it would take too long to convert each new message into electronic form and that some messages would be too long to be stored. Flowers understood and accepted this point, quickly revising his proposal to one where only the chi streams need be stored electronically. Within two weeks Newman wrote a handwritten note to Travis stating:[17] "For the more ambitious machine they [the Post Office] now propose to use tape for the message and valves only for the fixed wheels. This does away with the main objection to their first scheme (lack of flexibility for us)."

Nevertheless, Bletchley Park decided to back the two-tape machine rather than gambling on Flowers's design. By June, the first two-tape machine – featuring a combining unit that Flowers and his team developed as per the original request – was delivered to Bletchley Park.

---

[16] James A Reeds, Whitfield Diffie, J V Field (eds.), *Breaking Teleprinter Ciphers at Bletchley Park* (Wiley IEEE Press, ISBN 978-0470465837), Appendix D: Initial conception of Colossus, p535. Also available at 'GCHQ celebrates 80 years of Colossus', **gchq.gov.uk/news/colossus-80**, 18 January 2024

[17] As above, p536

It was quite something to behold. So much so that it was named the 'Heath Robinson' by the Wrens who operated it due to its elaborate, cumbersome design (William Heath Robinson was a famous cartoonist of the time who created comically complicated machines to perform a simple job; cartoonist Rube Goldberg earned similar fame in the USA). It worked, but only after weeks of effort by Newmanry section members Jack Good and Donald Michie.

Bletchley Park ordered more, and an improved, finessed version arrived in November. This slicker machine lost the 'Heath' and was simply called a Robinson. And Flowers's initial reaction that the design wouldn't work proved incorrect: in fact, their flexibility meant they were still being built and improved upon until the end of the war. The later versions were even christened Super Robinsons.

While Bletchley Park wouldn't commit money to the Colossus, they also said that the Post Office could go ahead with the design at its own risk and cost. Fortunately, Flowers had the backing of Gordon Radley, who understood the importance of the work and gave his chief engineer access to all the materials and people he required.

"I knew what it would do," said Flowers. "I drew a lot of square boxes, with lines joining up to these, and I specified what each box had to do." For example, how each box would simulate one of the code wheels. Flowers had to work alone until others could be let into the secret, but was soon joined by talented colleagues.

"Sid Broadhurst was a master of system design, and of the logic and timing of complicated relay circuits," wrote Harry Fensom, another member of the team at Dollis Hill, in his article 'How Colossus was Built and Operated – One of its Engineers Reveals its Secrets'.[18] "Bill Chandler was the expert on digital valve design, and could produce a system that worked straight away when implemented from the drawing board."

In fact, while some histories simplify the computer's story and say that Tommy Flowers "built" the Colossus, he had a large team of engineers working alongside him in the lab, plus the considerable manufacturing power of the Post Office to draw upon. Fensom, who contributed to later Colossi's design and worked on-site at Bletchley Park once the machines were installed, added that other engineers helped

---

[18] Harry Fensom, 'How Colossus was Built and Operated – One of its Engineers Reveals its Secrets', in Jack Copeland (ed.), *Colossus: The Secrets of Bletchley Park's Codebreaking Computers*, Chapter 24, p301

The 'set total' switch panel on Colossus
Image from General Report on Tunny: With Emphasis on Statistical Methods (1945), UK Public Record Office, Public Domain

draw "the detailed layouts and thought through the fine print of the design" while "dedicated wiremen" soldered it all together.

Flowers started designing the first Colossus in February 1943, with the machine delivered to Bletchley on 18 January 1944 and operational on 5 February.[19] So, a year later, just as Flowers had predicted[20]. It filled a room, more than justifying its name. While the machine was loud – thanks to the use of many electromechanical parts, not the valves – it worked from the get-go. Not only was it quicker than the existing Robinson machines, when they ran a message twice they obtained the same result. This wasn't always the case with the Robinsons, even the improved versions.

In his seminal report 'The Colossus', computer historian Brian Randell quotes Flowers directly: "They just couldn't believe it when we brought this string and sealing wax sort of thing in and it actually did a job. They were on their beam ends at the time, Robinson just hadn't got enough output, they wouldn't go fast enough,

---

[19] This is according to Gordon Radley's war diary, available for download at bt.com/about/bt/our-history/bt-archives/dollis-hill. See War Diary Vol 2, p227

[20] Some histories state that it was running at Bletchley in December 1943, and this is also what Flowers says in his 1998 interview. However, we know from official records that it was only installed and working at Bletchley Park in February 1944. Copeland suggests that December 1943 is likely to be the date when Flowers successfully completed a test run on the machine, which was built at Dollis Hill.

Chapter 4: Colossus   75

and suddenly this bit of string and sealing wax, in about ten minutes … and then they started to take notice!"[21]

Gordon Radley's wartime diary[22] is more measured in its analysis, but still the delight at its success rings through. "Since February 22nd Colossus has been in full daily operation for 16 hours per day, and, although the operating staff are not yet sufficiently familiar with it to utilise it to the best advantage, it has dealt with eight messages in eleven hours. For the remaining eight hours of its day, Colossus is turned over for the attention of Research Staff who are endeavouring to graft on to it equipment for providing still further facilities."

Flowers and Radley expected Bletchley Park to immediately place orders for more machines, so were surprised when they didn't hear anything. Still, they decided to press ahead with the manufacture of the more time-consuming components so that they would be ready when the call came.

It's a good thing they had a head start, because the following month they were visited by "some man in uniform, straight from the War Cabinet, [who said] that we had to make twelve of these machines by June the 1st," remembered Flowers. By then it was obvious to all that it was only a matter of time before the Allies invaded Europe, and their visitor's next statement made Flowers suspect that his machines might just play an important role. "We were told that unless you can make these machines and have them working by June the 1st then it would be too late."

While the Post Office couldn't build a dozen machines in such a short time, Flowers promised that the first new machine would be working by 1 June with the next two following in July and August. So began an intense four months of effort to build the Colossus II, this time with 2400 valves. "We worked day and night, six days a week," said Flowers. "We worked ourselves almost to exhaustion."

That work continued right up until the last moment. Only at 1am on the morning of 1 June 1944 did Flowers and his team work out that the reason the Colossus II obstinately refused to work was due to misbehaving valves. This was a problem best left to valve expert William Chandler, so they left him there. "And then Flowers came back the next morning and Colossus II was buzzing away like a top," says Copeland.

---

[21] Brian Randell, 'The Colossus', University of Newcastle upon Tyne, 1976, **homepages.cs.ncl.ac.uk/brian.randell/Papers-Books/133.pdf**

[22] Gordon Radley's war diary, **bt.com/about/bt/our-history/bt-archives/dollis-hill**, p227

"Chandler was looking very pleased with himself, but also standing in a pool of water because the radiator had leaked during the night." This, incidentally, is why some reports incorrectly state that the Colossus was water-cooled; ignore all such articles.

"Between [5am] and 8pm this evening, Colossus II was running continuously and has very successfully dealt with no less than 20 separate tasks," wrote Radley in his diary that day. "This is roughly equivalent to a full 24-hour output from all the other machines at present operated by the Group concerned, and has given very great satisfaction to Professor Newman and Commander Travis."

The Colossus II was so quick thanks to the advent of what can best be described as parallel processing. This involved having the equivalent of five processors working on the same tape, effectively bringing the speed up to 25,000 characters per second. "It meant more gates and counters, of course," wrote Fensom,[23] "but we didn't mind a thousand extra valves."

Nor did Bletchley Park mind throwing hundreds of women at the problem. For while the Colossi revealed the settings for the five chi-wheels, there was still much work to be done before a message could be decrypted, and much of this burden fell to Wrens and women from the army's Auxiliary Territorial Service.

"There were so many messages being sent in, when one didn't come good you were sent another one," Cora Jarman told Tessa Dunlop, author of *The Bletchley Girls*,[24] of her time in the Newmanry. "Checking checking. It was very boring." The tedium wasn't helped by the fact that the women weren't privy to the methods, they were merely told what to look for.

After a Colossus had successfully revealed a likely set of starting positions for the chi-wheels, the message needed to be 'de-chi'ed'. This meant processing it through a machine to remove the chi encryption, which one of the women would do. The de-chi'ed message tape would then be taken to the Testery, where the code breakers would set to work. When a complete set of wheel positions arrived, it was women who would set up a Tunny machine and check the output.

So it was hard work for everyone involved, but also vital work. According to legend, on 5 June 1944, General Eisenhower was handed a decrypt fresh from Bletchley Park.

---

[23] Harry Fensom, 'How Colossus was Built and Operated – One of its Engineers Reveals its Secrets', in Jack Copeland (ed.), *Colossus: The Secrets of Bletchley Park's Codebreaking Computers*, Chapter 24, p301
[24] Tessa Dunlop, *The Bletchley Girls* (Hodder & Stoughton, 2015, ISBN 978-1444795745), p124

He read it, then declared: "We go tomorrow." This may be too neat a story, but it's certainly known that Eisenhower was a regular visitor to Bletchley Park in the run-up to D-Day and an admirer of its work.

We should not oversimplify history to say that the Colossi alone won or shortened the war. For one, we can't know if Eisenhower would have still invaded the following day if he hadn't read that message. But giving all the credit to the computer ignores the months of work done by a network of double agents who had been feeding the Nazis with misinformation.[25] Not to mention other ruses, such as creating fake barracks near Dover, or dropping aluminium foil to create the illusion of incoming planes on German radar. All these deceptions were to convince the German military that the Allies were going to land in Pas-de-Calais rather than their real target of Normandy.

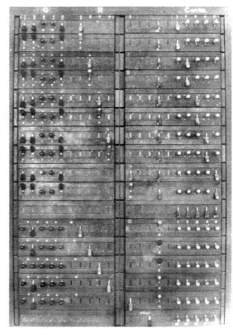

Colossus K2 switch panel showing switches for specifying the algorithm (left) and the counters to be selected (right)
Image: UK Public Record Office, Public Domain

The deception worked. The Nazis held back reinforcements from Normandy, allowing the Allies to establish a strong foothold there. Over the coming months, as the Allies battled their way across Europe, more Colossi were installed. By the end of 1944, seven whirred away, spread across two vast, bomb-proof concrete buildings. Three more Colossi arrived in 1945. One mathematician might be in charge of two Colossi, and they would give Wrens plugging diagrams to follow. If something went wrong, Post Office engineers were on hand to fix the problem.

---

[25] This was Operation Fortitude South, part of a wider plan called Operation Bodyguard, geared at fooling the German High Command into expecting an attack at Pas-de-Calais.

"I saw this astonishing machine the size of a room," said Joanna Chorley.[26] "It was ticking away, and the tapes were going round and all the valves, and I thought what an amazing machine... Like magic and science combined!"

What's more, the machines became more magical over time. For example, Bill Tutte created a technique called rectangling that could determine the wheel patterns. However, it required a quite legendary amount of effort by hand and one tiny mistake could ruin the calculations. Donald Michie, a talented member of the Testery, came up with a proposal that meant a Colossus could become a high-speed rectangling machine through the addition of a "rectangling gadget".[27] A good thing, as the Germans started to change the wheel patterns on a daily basis as they became more concerned about security.

"We were in no doubt about how important [Colossus] was," said Captain Jerry Roberts in 2010,[28] a member of the Testery. "We saw a number of messages signed by Adolf Hitler himself." The operation became slick, with Roberts adding that "I suspect sometimes we genuinely saw the messages before the blessed Germans."

The day they installed Colossus II marked the moment when Flowers moved away from day-to-day charge of the computers, as he took on a different project to design a computer to help target anti-aircraft guns. In this way, unknown to both men, Flowers and George Stibitz's war careers echoed each other (see Chapter 3 for more on Stibitz). It made sense to employ Flowers's talents elsewhere, as the template for each new machine was now established and his skills were too valuable to be wasted.[29] From this point on, Allen Coombs took control.

For the rest of the war in Europe, the Colossi continued to break codes, with the tenth and final operational machine installed in April 1945. A month later, the teleprinters relaying the Tunny messages would fall silent – an eerie experience for those working in the huts – with the final message sent shortly before VE Day on 8 May 1945, when Nazi Germany unconditionally surrendered to the Allied forces.

---

[26] Tessa Dunlop, *The Bletchley Girls*, p125
[27] Jack Copeland (ed.), *Colossus: The Secrets of Bletchley Park's Codebreaking Computers*, p244
[28] 'Bletchley's code-cracking Colossus', BBC News, 2 February 2010, **news.bbc.co.uk/1/hi/technology/8492762.stm**
[29] After the war, Flowers stayed in position at Dollis Hill and helped the General Post Office in its push to a modern telephone network. His only computing involvement was in the early stages of the Pilot ACE, as described in Chapter 10, which eventually resulted in the MOSAIC. This was designed by Allen Coombs and William Chandler, who worked under Flowers, and is considered the truest implementation of Turing's ACE design. MOSAIC is said to have provided good service to the Ministry of Supply from 1955 until its retirement in the early 1960s.

A lack of new messages didn't spell the end for the Colossi. There was still much to be learnt from the backlog of messages still to be deciphered, including clues to the identities of British agents giving material to the Soviet Union. For while the USSR was an ally of sorts, it was always an uneasy relationship.

Moscow must have known about the British success decrypting Tunny, because British civil servant John Cairncross – later unmasked as the fifth member of the infamous Cambridge spy ring that included Kim Philby and Guy Burgess – gave them decrypts that he smuggled out of Bletchley Park by stuffing them down his trousers. But the USSR clearly didn't realise how comprehensively Tunny had been broken, as it grabbed the Nazi cipher machines and used them (with enhancements) to encrypt messages after the war.

With this in mind, it's easy to understand why the British government wanted to keep its secrets safe. We don't know if Cairncross knew of the Colossi's existence, such was the level of secrecy within Bletchley Park, and so it could be that the Soviets had no idea about the existence of Britain's high-speed computers.

However, there was no need to keep all ten active Colossi in service, so the command came to dismantle them. In a 2012 short film to celebrate the 70th anniversary of Colossus's birth,[30] Google interviewed engineers and wirers who worked on the machines. Their most poignant remarks came when describing the Colossi's demise.

"When we had the order to dismantle all the equipment, the idea was to smash everything. Smash all the valves and everything else," remembered Roy Robinson.[31] "Some of [the Colossi] were thrown down coal mines," added Albert Bareham. "The whole piece of equipment was dumped down a disused coal mine and things like that."

Even 70 years later, you can hear the disappointment in the voice of Margaret Bullen (née Boulton), who like other women "with nimbler fingers" helped with wiring and soldering, when she remembers the end of her time at Bletchley Park. "We carried on … until the war was over, and immediately we were told we could go," she said. "You felt cheated in a way that you'd worked so hard on it and that it was going to be no more."

---

[30] 'Colossus: Creating a Giant', **youtu.be/knXWMjIA59c**, 8 March 2012
[31] While this may have been the fate of some of the Colossi at Bletchley Park, we know that a truck's worth of components were shipped to Manchester at the request of Max Newman – as we cover in the story of the Manchester Baby in Chapter 7.

None of them knew at the time that at least two Colossi survived the massacre. These were quietly transported to the new headquarters of CG&GS — renamed Government Communications Headquarters, GCHQ, the following year — in northwest London. The Colossi then moved with GCHQ to Cheltenham in 1951, where they continued to provide valuable code-breaking service for another decade.

Colossus selection panel with far tape selected
Image: UK Public Record Office, Public Domain

"I worked as an engineer on Colossus for a year during the 1960s," wrote Bill Marshall, a former GCHQ engineer, in an article on GCHQ's website to mark Colossus's 80th anniversary.[32] "I was told very little about the machine I was working on – what the machine was actually doing was not for me to know. My job was to repair it as necessary, using just a few circuit diagrams and no detailed user handbook."

It's likely that Marshall was one of the final engineers to tend to the machines, with the same article stating that Colossi stayed in use "until the early 1960s". Despite the computers' retirement, the government refused to declassify the Colossus project. It was only through the efforts of historians, particularly Brian Randell, that more information came to light in the mid-1970s. It took another quarter of a century for the government to release a 500-page report written by members of the Bletchley Park team, 'General Report on Tunny',[33] that filled in many of the missing details.

"ENIAC demonstrated very publicly the viability of using large numbers of electronic tubes," said Randell in an interview now found on YouTube,[34] "just as Colossus had done on a smaller scale, but privately. Colossus, I think, had an effect

---

[32] 'GCHQ celebrates 80 years of Colossus', **gchq.gov.uk/news/colossus-80**, 18 January 2024
[33] This can be read in full here: **alanturing.net/turing_archive/archive/index/tunnyreportindex.html**
[34] The Computer Pioneers: Interview footage of Brian Randell, **youtu.be/9HFyrJq3e1E**

within Britain, but a rather limited effect, whereas ENIAC brought 'giant electronic brain' onto front pages."

Which brings us to the big question of the impact of Colossus. Randell doubted that it had much effect on Alan Turing – "One of his major characteristics is his wish to start everything from scratch" – but there is little doubt that it inspired Newman to start work on a computer at the University of Manchester. This would give birth to the Manchester Baby, subject of Chapter 7.

Even though some in American intelligence knew about the Colossus, we can also state with some confidence that it had no direct impact on the growth of computing in America. They had the ENIAC and detailed plans for the EDVAC, and unlike the reticent British the Americans were happy to share their wartime advancements.

This does not mean the Colossus should be ignored in computing history, as it marked many firsts. The Colossus I was the first computer to use electronic binary circuitry on a large scale. The first to use a clock pulse to synchronise operations. And while it was a special-purpose computer, it and its successors were programmable – to an extent.

Timing naturally becomes important when claiming world firsts, so it should be noted that the first Colossus was operational in February 1944, pre-dating the ENIAC by almost two years. This was why, when Randell – with the assistance of Coombs – gave a guarded presentation on the Colossus to the International Research Conference on the History of Computing, Los Alamos, in 1976 it caused quite a stir.

Bob Bemer, who helped develop the ASCII character code, wrote a short but vivid article describing the proceedings.[35] "My decision to keep everyone in view paid off. I looked at Mauchly [John Mauchly of ENIAC fame], who had thought up until that moment that he was involved in inventing the world's first electronic computer," he wrote. "I have heard the expression many times about jaws dropping, but I had really never seen it happen before. And Zuse – with a facial expression that could have been anguish."

All of which begs the question, what might have been different if the Colossi's creators hadn't been sworn to secrecy? Could it have given extra momentum to Britain's efforts? Might it have changed computing history?

---

[35] A copy of the document can be found at **rpimag.co/colossusbemer**, and it is quoted extensively in 'The First Public Discussion of the Secret Colossus Project' by Michael Williams, in *IEEE Annals of the History of Computing*, Vol 40, Issue 1, Jan–Mar 2018

What we know for certain is that Tommy Flowers, and others involved in the creation of the Colossus, never received the recognition they deserved in their lifetimes. Flowers was given a one-off gift of £1000 (equivalent to roughly £35,000 in 2025) after the war, but is said to have shared that with others involved in the project.[36] He also collected an MBE in 1943 for his early war efforts, before he even started work on the first Colossus. Standing for Member of the British Empire, this is a national honour, but among the lowest-ranking.

It was only through the effort of Randell that Flowers gained greater recognition in later life. This included an honorary degree from Newcastle University (where Randell was then a professor and remains a professor emeritus) in 1977,[37] but when asked how he felt about recently winning the Charles Babbage award in the late 1990s Flowers's response was muted. "I appreciate it," he said, "but it isn't the same as at the time. It would have been fine if it had happened in 1946, but in 1998 it doesn't really mean much. I'm quite honoured but I'm not over-enthusiastic."

Perhaps he would have appreciated the BBC documentary *Code-breakers: Bletchley Park's Lost Heroes*, which shone a light on the efforts of Tutte and Flowers in breaking Tunny. Broadcast in 2011, the hour-long film is still available to watch online.[38]

More accolades also followed Flowers's death. There is now a blue plaque commemorating the Dollis Hill Research Station, while British Telecom – the organisation that inherited the Post Office's national telephone duties in 1981 – created a statue of Flowers at its research facility in 2013. If anything, Tommy Flowers's fame and repute continues to grow, with streets and even a pub now named after him.[39]

Colossus itself lives on in two forms. The first is a fully working recreation of Colossus II, created through the hard work of dozens of engineers but the dogged

---

[36] Some online articles state that Flowers paid for parts of the Colossus himself, but this is unlikely to be true. "I think Flowers meant he paid the personal cost," says Copeland, referring to an interview where Flowers mentioned being "out of pocket" for Colossus. "He's in the Dollis Hill Research Station. They've got relays and vacuum tubes piled up behind every door. What would he have to pay for?"

[37] Ignore sources that suggest it was given in 1973. The date 1977 is taken from Newcastle University's official listing, found at **ncl.ac.uk/congregations/honorary/honorary-graduates/**

[38] It can be found directly at **clp.bbcrewind.co.uk/7fd3fb55e462db0867b183729c5ed27c**, but it's easier to visit **clp.bbcrewind.co.uk** and search for 'Timewatch code-breakers'.

[39] Tommy Flowers Drive is in Kesgrave, near the BT Laboratories, while the Tommy Flowers Mews is in North London. The community-run Tommy Flowers Pub is based in Poplar, London, where Flowers was born.

determination of one man: Tony Sale. "In 1991, some colleagues and I started the campaign to save Bletchley Park from demolition by property developers," he wrote.[40] "At the same time I was working at the Science Museum in London restoring some early British computers. I believed it would be possible to rebuild Colossus, but nobody else believed me."

With the help of still-classified documents, circuit diagrams that engineers had illegally kept, and Arnold Lynch, one of the men behind the photoelectric reader, Sale first recreated the design. Then, this time helped by a mix of "current and retired Post Office and radio engineers", they built a limited version of the first Colossus – switching it on in time for the 50th anniversary of the ENIAC in 1996. A ceremony Tommy Flowers attended.

By 2003, after what Sale described as "over 6000 man-days of effort", the replica of the first Colossus was working almost exactly as designed. But, not content with this, the team decided to "upgrade" to Colossus II. Their self-imposed deadline was 1 June 2004, exactly 60 years after the second computer became operational. This time, there would be no misbehaving valves. "[On] Thursday 20th May I filmed Colossus Mk 2 setting all five K wheels [psi-wheels] on the Bream cipher text and after editing, this video was shown to 120 people at our commemoration event at the Science Museum in London on 1st June 2004," wrote Sale.[41]

The rebuilt Colossus II remains on show at The National Museum of Computing, which is on the same site as Bletchley Park but is run separately.[42] The museum also houses what it claims is the world's largest collection of historic computers and, unlike in most museums, visitors are encouraged to get hands-on with them. The Colossus is very much 'look but don't touch', unless you're one of the volunteers who keep it going.

The Colossus is also kept alive by a virtual version – search for 'virtual Colossus' online – where you can see a 3D representation of the computer and even set it to work. While the original program was written by Tony Sale, it was Martin Gillow who recreated it in 3D and brought it up to date.

---

[40] Tony Sale, 'The Colossus Rebuild', in Jack Copeland (ed.), *Colossus: The Secrets of Bletchley Park's Codebreaking Computers*, Chapter 12, p150

[41] 'Rebuilding Colossus' by Tony Sale, The National Museum of Computing website, undated, **tnmoc.org/rebuilding-colossus**

[42] Visit Bletchley Park's website at **bletchleypark.org.uk** and The National Museum of Computing's website at **tnmoc.org**

All of which means that despite the best efforts of the British government, the very modern legend of Colossus lives on.

Left side of the Harvard Mark I displayed in the Cabot Science Building, Harvard University, 2005

Image: Waldir Pimenta, CC BY-SA 3.0

Harvard Mark I

The first in a series of pioneering electromechanical machines

Howard Hathaway Aiken was a remarkable man. Not merely because he was the driving force behind the world's first genuinely useful computer, but because there were two starkly contrasting sides to his personality: on one side, the fierce, brisk and often unforgiving head of department. A man who ruled the Harvard Computational Laboratory with an iron fist. On the other, a thoughtful, loyal, entertaining friend and a loving son devoted to his mother.

That sense of loyalty came to the fore in 1912, when a 12-year-old Aiken confronted his drunkard father Daniel who was – not for the first time – physically abusing his wife, Margaret Emily Mierisch Aiken. According to Bernard Cohen's account in *Howard Aiken: Portrait of a Computer Pioneer*,[1] the young Howard "grabbed a fireplace poker and drove his father out of the house".

That was the last they saw of Daniel and, more crucially, his family money: Howard Aiken's paternal grandparents, who had helped to support the young family, immediately cut funds to mother and son. An only child, this meant Aiken became financially responsible for himself, his mother, and even her parents – who lived with them – once he completed eighth grade at the age of 14.

At this point Aiken's future, and the creation of the Mark I, was hanging by fate's thread, only for a teacher to intervene. Aiken was a brilliant student, but money was more immediately important to the family than schooling. The teacher pleaded with Margaret to change her mind, but the rent needed to be paid, food put on the table. She needed her only child to earn money.

Fortunately, Aiken's teacher (whose identity has now been lost) was not merely tenacious but also imaginative. He, or she, found the young scholar a night job at the Indianapolis Light and Heat Company as an electrician's helper.[2] Once his shift ended, the money earned, Aiken could head to school.

To pause for a moment, this means that a teenage Howard Aiken was pulling all-night shifts, then studying, leaving little time for rest. Although Aiken claimed that some of his nights were spent manning the switchboard, allowing him to grab a few hours' sleep on occasion. Of all the people covered in this book, many of whom lived privileged lives, there's little doubt that Aiken's teenage years were the toughest.

---

[1] I Bernard Cohen, *Howard Aiken: Portrait of a Computer Pioneer* (The MIT Press, 1999, ISBN 978-0262032629), p9
[2] As above

Despite his challenging schedule, Aiken worked hard at school and gained not only his education but also support from Indianapolis's Superintendent of Public Instruction, Milo Stewart. In a 1973 interview with Henry Tropp and Bernard Cohen, Aiken explained that he went to the University of Wisconsin because Stewart sent letters to public utilities "in every Midwestern University town",[3] and only the Madison Gas and Electric Company offered him a job.

Aiken was now 19 years old, studying electrical engineering, and would emerge with a Bachelor of Science degree in 1923. But financial factors would continue to slow his path. Rather than stay on at college, he used his degree to secure a better job at the Gas and Electric company and would only re-enter academia almost a decade later.

Aiken spent five of those years as an engineer at the Westinghouse Electrical and Manufacturing Company in Chicago, where he earned good money but found that his intellectual itch was not being scratched. He resigned from his job and signed up at the city's university, but only spent two quarters of the academic year there "upon discovering that the faculty were bootlegging grades at the behest of the newly appointed president".[4]

Spurred on by a professor at Chicago, he switched to Harvard, a centre of excellence for physics. By this time, autumn 1933, Aiken was in his mid-30s and significantly older than most graduate students. Ronald King – who joined Harvard at a similar time to Aiken and would go on to be a professor of applied physics there for almost 30 years – later told Cohen that this tall young man, who was certainly aware of his own brilliance, did not always get on with the Physics Department.

To such an extent that he might have been "kicked out," said King,[5] were it not for the intervention of Professor Emory Leon Chaffee, who headed the Communication Engineering group. Chaffee essentially took Aiken under his wing, and decided that his bright graduate student should work on space charge as his PhD thesis.

As was the case for Professor Atanasoff with the ABC (see Chapter 1), Aiken saw the need for faster and more reliable calculators during his research. Aiken said that "the object of the thesis almost became solving nonlinear equations," and doing so was both tedious and time-consuming on existing desk calculators. He added: "it

---

[3] Howard Aiken oral history interview with Henry Tropp and I B Cohen, 26 February 1973, Smithsonian National Museum of American History, **rpimag.co/aikeninterview**, p119
[4] As above, p120
[5] I Bernard Cohen, *Howard Aiken: Portrait of a Computer Pioneer*, p22

became apparent to me at once that this could be mechanised and programmed and that an individual didn't have to do this." [6]

Aiken set to work to devise such a machine, finding inspiration in a book entitled *Modern Instruments and Methods of Calculations: A Handbook of The Napier Tercentenary Exhibition*. This exhibition took place in Edinburgh in July 1914, 300 years after the Scottish mathematician John Napier published *A Description of the Admirable Table of Logarithms*. While the 'handbook' was merely designed to accompany the exhibition, its 343 pages would have given Aiken a comprehensive oversight of calculating machines' development from Napier's bones to slide rules to early 20th century innovations such as the Monarch-Wahl Adding and Subtracting Typewriter: "You type the amounts – the Monarch-Wahl gives you the total," read the company's newspaper adverts.[7]

The British Library still holds two copies of the handbook: one sits undisturbed in a low oxygen chamber, but a second is available for visiting Readers.[8] Four pages in particular must surely have grabbed Aiken's attention, laying out the key principles behind Babbage's Analytical Engine. The description covered his concept of the "mill", "store" and "Jacquard apparatus." [9] These are signature inclusions, albeit with different names, in the IBM Automatic Sequence Controlled Calculator, or Harvard Mark I as it came to be known.

We can't know exactly how influential this handbook was, because Aiken's own account of the Mark I's creation appears to have slipped into legend. In 1973, shortly before his death, Aiken said that "the first time he ever heard of Babbage" was when he discovered that two wheels from Babbage's Difference Engine were sitting in "the attic of the old research laboratories" at Harvard University.[10]

---

[6] Howard Aiken oral history interview with Henry Tropp and I B Cohen, 26 February 1973, Smithsonian National Museum of American History, **rpimag.co/aikeninterview**, p2

[7] *The Daily Telegraph*, 19 November 1913, p13

[8] Anyone over the age of 18 can become a Reader at the British Library. You apply on-site, explain the project you're working on, and are given a badge entitling you to enter one of the many Reading Rooms. From here, you request one of the 170 million items stored across its London and Yorkshire sites. Unlike a regular library, you cannot take any borrowed books out of the Reading Rooms.

[9] PE Ludgate, 'Automatic Calculating Machines', in E M Horsburgh (ed.), *Modern Instruments and Methods of Calculations: A Handbook of The Napier Tercentenary Exhibition* (The Royal Society of Edinburgh, 1914), p124

[10] Howard Aiken oral history interview with Henry Tropp and I B Cohen, 26 February 1973, Smithsonian National Museum of American History, **rpimag.co/aikeninterview**, p4

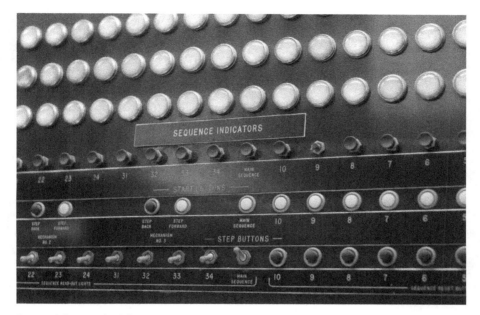

Sequence indicators and switches
Image: Rocky Acosta, CC BY 3.0

He told the story to illustrate the university's resistance to his computing project. "The faculty had rather limited enthusiasm about what I wanted to do, if not almost downright antagonism," said Aiken.[11] He added that one member of the laboratory "couldn't see why in the world I wanted to do anything like this in the Physics Laboratory, because we already had such a machine and nobody ever used it."

Aiken set out to find this machine, only to discover a pair of forgotten wheels that had been presented to Harvard by Charles Babbage's son. The fact these were gathering dust in an attic reflects how distant a memory Babbage had become, which is perhaps surprising when you consider the impact he had in 19th century Britain. (See the Introduction for more on the Babbage story.)

Now we fast-forward to 22 April 1937: the date on which Aiken presented his plans to Monroe Calculating Machine Company, then a leading manufacturer of desk calculators.

---

[11] As above, p3

It's crucial to note that, unlike other pioneers such as Atanasoff, Mauchly, and Zuse, Aiken's proposals focused on what he wanted his invention to do rather than the mechanics: calculations such as tabulating Bessel functions, a time-consuming task that would give the Mark I its nickname of Bessy. He essentially presented an architecture that a manufacturer would then convert into a functioning machine using its own expertise. Aiken later claimed that this meant that if Monroe had decided to partner with him, the computer would have been made out of mechanical parts. If RCA had been interested, it might have been electronic. And it follows that it was made out of electromechanical, tabulating machine parts because IBM was willing to pay the bill.

At that time, Monroe was a big name in desk calculators, so it was natural for Aiken to approach the firm. He met with George Chase, director of research, who shared the details of the meeting in a slide show detailing the history of mechanical computers in 1952 – published in the *IEEE Annals of the History of Computing*.[12]

"He told me certain branches of science had reached a barrier that could not be passed until means could be found to solve mathematical problems too large to be undertaken with the then-known computing equipment," wrote Chase. "He outlined to me the components of a machine that would solve those problems."

Chase concluded: "What he had in mind at that time was the construction of an electromechanical machine, but the plan he outlined was not restricted to any specific type of mechanism; it embraced a broad coordination of components that could be resolved by various constructive mediums. I knew then that the second era of development of computing machinery had started."

Unfortunately for Chase, and Monroe, after many months of going back and forth he could not persuade the company's board to pay the high cost of the machine. It must have been with great reluctance that Chase suggested to Aiken that he contact IBM instead.

Automatic computing wasn't entirely new ground for International Business Machines. In 1936, it had built a Calculation Control Switch designed by a Columbia University Astronomy Department professor called Wallace Eckert (no relation to Presper Eckert of ENIAC fame). Eckert would eventually head up IBM's 'super

---

[12] George C Chase, 'History of Mechanical Computing Machinery', in *Annals of the History of Computing*, Vol 2, No 3, July 1980

Rotary switches used to enter program data constants
Image: Arnold Reinhold, CC BY-SA 3.0

calculator' project in 1947,[13] which became the Selective Sequence Electronic Calculator (SSEC). Eckert's 1936 switch made astronomical calculations much quicker by using punched cards to send commands to IBM's existing Type 601 electric multiplier.

Aiken's proposal was an order of magnitude more complicated than the Calculation Control Switch, however, and also far more expensive. The initial estimate rapidly grew from around $15,000 to $100,000 and, forming a precedent for IT projects that persists to this day, its cost would more than double again. Despite this, IBM would eventually donate the Mark I and an additional $100,000 to Harvard as a contribution to its running costs. In return, IBM owned all the patents resulting from its construction.

One of the reasons Thomas Watson Sr, IBM's chairman, was willing to spend the money was because he wanted to replicate the strong links with Harvard that IBM

---

[13] This project would lead to the Selective Sequence Electronic Calculator (SSEC) – source: Emerson W Pugh, *Memories That Shaped an Industry* (MIT Press, 2000, ISBN 978-0262661676), p7

already had with Columbia University. But, as we shall see later in this chapter, the course of history didn't run so smooth.[14]

Another reason for IBM saying yes is because Aiken found a backer for the project in James Bryce, who was both the company's chief engineer and Watson's most trusted technical adviser. According to IBM historian Emerson Hugh, Bryce "had no trouble convincing Watson of the merits of the project."[15]

Still, progress was slow. We know Bryce and Aiken first met in November 1937, after which Bryce arranged for Aiken to visit Eckert at Columbia to see if the IBM calculator could achieve all Aiken wanted. The answer: a firm no. Bryce decided Aiken needed a better idea of what IBM's equipment could do, so next arranged for him to attend official IBM training classes and even accompany service engineers on jobs.

But Aiken's most intense inauguration into IBM world came when he spent four days at IBM's manufacturing facility in Endicott, New York, early in 1938. Here, he worked closely with the two engineers who would essentially build the Mark I: Frank Hamilton and Benjamin Durfee. Over the course of those four days, the three men attempted to set out the requirements of the machine, down to what could be created from existing IBM parts and what would need to be designed from scratch.

At this point, both parties believed that the computer would be built at Harvard with IBM's assistance as required. As such, Aiken still needed to gain the backing of his university. Although Harry Mimno, a close colleague of Aiken's at Harvard, would later describe his role in the creation of the Mark I as that of "an enthusiastic and fascinated listener",[16] he was the one who wrote a compelling report to Harvard's then-President, James Conant, a matter of days after Aiken returned from Endicott. In this report, he emphasised the minimal cost to Harvard – a "few thousand dollars," he wrote – and the immense return they would get in a calculator that could be put to use in a wide variety of academic research projects. His examples, presciently, included atomic physics.

Even without a formal agreement in place, IBM's Bryce kept the project moving in the right direction. Described by Aiken as "an astute inventor" who was "worth

---

[14] Emerson W Pugh, *Building IBM: Shaping an Industry and its Technology* (MIT Press, 1995, ISBN 978-0262161473), pp73-76
[15] As above, p73
[16] I Bernard Cohen, *Howard Aiken: Portrait of a Computer Pioneer*, p 37

all other IBM people put together",[17] it was Bryce who sent Hamilton a layout of the Mark I's panels – having devised them in partnership with Aiken. In May 1938, IBM formally put money behind the project by giving Hamilton the funds to start work on a design on how the components would fit together.

While history knows the Mark I's importance, at this point IBM did not. Thomas Watson Sr never envisaged that it would be the beginning of a new line of products. He thought that it would be merely a scientific and academic partnership – much like the calculator IBM had created for Eckert at Columbia. As such, the project lived on the bottom rung of the Endicott team's priorities, and Hamilton didn't start work on the design until August.

It took a further eight months – during which Aiken made numerous visits to Endicott to provide information and chivvy the project along – before Harvard and IBM signed the final agreement to construct the machine. During the previous months, it had become obvious that Harvard was in no position to build the Mark I, so that duty fell to IBM. In particular to Hamilton and Durfee, with the ultimate responsibility falling to chief engineer Clair Lake.

Although Aiken was later dismissive of Lake's contributions – "Lake contributed absolutely nothing,"[18] he said in 1973 – there is plenty of evidence to suggest otherwise. According to Hamilton, Lake designed a new relay, and co-designed (with Hamilton) the punch for preparing the sequence tape on which instructions were passed to the computer. Ultimately, he was also in charge of the budget, reining back Aiken's more extravagant requests whilst delivering practical solutions.

Nor could the Mark I have been built without Durfee. His role was to devise the circuitry, and as the computer moved further away from existing IBM components the more work he had to do. A modest man, he perhaps undersold his talents: in an interview with IBM historians, he said he "had never been too good at mathematics,"[19] yet he covered an "enormous piece of drawing paper… with pencil arithmetic on all sides" to prove to himself that Aiken's suggested method of division would work.[20]

---

[17] Howard Aiken oral history interview with Henry Tropp and I B Cohen, 26 February 1973, Smithsonian National Museum of American History, **rpimag.co/aikeninterview**, p23
[18] As above
[19] I Bernard Cohen, *Howard Aiken: Portrait of a Computer Pioneer*, p 74
[20] Howard Aiken oral history interview with Henry Tropp and I B Cohen, 26 February 1973, Smithsonian National Museum of American History, **rpimag.co/aikeninterview**, p24

Input/output and control readers
Image: Waldir Pimenta, CC BY-SA 3.0

Aiken continued to visit Endicott regularly, both to check on progress and to agree on elements of the design with Lake and Hamilton. By this point, Bryce had moved onto other projects, leaving Aiken as the theoretician while the IBM engineers interpreted his ideas, requests, and sketches into something that could be built without costs spiralling out of control. For example, Aiken wanted three multiply-divide devices, and that dropped down to one in the final machine. Similarly, he envisaged multiple sequence control units, and this again became one. Durfee estimated that if they had agreed to all Aiken's suggestions the Mark I would not only have been prohibitively expensive but also three times its ultimate size. And bear in mind the final machine was "longer than a diesel locomotive,"[21] according to The Boston Globe.

Aiken's frequent visits and detailed instructions kept coming until the spring of 1941, when he was called up to the US Navy. While Aiken said that he "still made

---

[21] Dorothy Wayman, 'Harvard Gets Huge Calculator', in *The Boston Globe*, 7 August 1944, p7

those trips to Endicott to plead: 'Come on, let's do a little more',"[22] during his time in service, he passed the responsibility of finishing the job to a Harvard graduate called Robert Campbell.

Unlike the ENIAC, which had been commissioned by the army to calculate missile trajectories, at this point the Mark I had no formal military connection. Nor did those close to it appreciate its potential applications in the war effort. So while some of IBM's factories were being used to create weapons of war, the company was also under pressure to convert existing machines into a form that could be used by the US military. It's little wonder that the experimental Mark I dropped down IBM's pecking order.

Despite Campbell doing his best to apply pressure on Hamilton and colleagues, he also had the distinct disadvantage of being a second-year graduate student rather than a professor. Plus, there were certain questions that IBM felt only Aiken could answer, but for chunks of time they simply couldn't contact him.

To an extent, then, it seems remarkable that the project wasn't quietly sidelined. Instead, IBM, Campbell, and Robert Hawkins – a graduate in 'Electrical Construction'[23] who had been working in Harvard's Physics laboratory but was sent to Endicott so he could help with development – kept things moving along as rapidly as they could. And at last, on 1 January 1943, almost six years after Aiken had first approached IBM, the Mark I solved its first 'real' problem.

It wasn't very exciting. According to Campbell's notes, the Mark I computed and graphed "the time required to build up the current in an inductive circuit based on an equation which required the machine to multiply, divide, add, subtract, compute logarithms and antilogarithms." And after this auspicious start? Silence. When Campbell sent details of the first successful program to Aiken, he heard not a word in reply. And we have no records of any other programs being run on the computer until December 1943.

We know that Aiken was spending at least some of his time lecturing on the subject of electricity to navy recruits at the Naval Mine Warfare School in Yorktown, Virginia. But there are also good reasons to suspect he had other more secretive duties. One rumour is that he personally decommissioned a German torpedo; another that

---

[22] Howard Aiken oral history interview with Henry Tropp and I B Cohen, 26 February 1973, Smithsonian National Museum of American History, **rpimag.co/aikeninterview**, p13

[23] Robert Hawkins oral history interview with William Aspray, 20 February 1984, Charles Babbage Institute, **rpimag.co/hawkinsinterview**, p3

he travelled to France by submarine, was smuggled into the country dressed as a peasant, all to inspect a secret piece of German weaponry. While we can never know for sure, it at least serves as a possible explanation for Aiken's lack of response to Campbell's message.

With no clear leadership from Harvard, and wartime distractions of their own, the Mark I was essentially forgotten for months. It could also be that the IBM engineers were dealing with the patents that had emerged during construction, and that IBM wished to complete that work before shipping the computer to Harvard as promised. There was a logistical side to this too, as dismantling and then rebuilding the Mark I would be a time-consuming task on its own.

What we know for sure is that finally, in October 1943, the President of IBM, Thomas Watson, sent a letter to Harvard's President James Conant inviting him for a demonstration of the completed machine later that month. Conant, however, was tied up with war duties of his own, with the visit delayed until December.

This, perhaps, was when the first seeds of discontent appeared in the Harvard-IBM relationship. From Watson's point of view, he had spent hundreds of thousands of dollars – in both parts and man-hours – creating this scientific calculator for Harvard. The least that Harvard could do was to appear grateful.

Now switch perspectives to Conant. He had never been enthusiastic about Aiken's project, once telling him that if he wanted to be promoted then he should stop wasting his time on mechanical computers. To Harvard's President, at this point, the Mark I must have seemed like a toy. He was so blasé about the project that during the ceremony to announce the machine, he reportedly asked who actually owned it.

Nevertheless, in December 1943 Conant made the trip to Endicott as per Watson's request. Aiken was also present for the demonstration and it appears to have rekindled his enthusiasm, if it ever went away. By spring the following year, he was desperate to be back in charge of the machine.

On Friday 5 May 1944, after two years away from his creation, Aiken returned to Harvard and took control of the project once more. From this point until the end of the war, it would largely be put to military use, starting with a project for the Navy's Bureau of Ships to classify steel – based on the properties of their known impurities – to determine where each batch should be best used. The mathematical problem, according to Cohen, required ten simultaneous linear equations to be solved and 'calculating multiple regression coefficients and multiple correlation coefficients'.[24] In short, a job worthy of a giant computer.

Fortunately, it wasn't long before his wish was granted. In his 1973 interview, Aiken described a "Naval Commander" calling the Navy school and said, "Why don't you come up here and run this computer?"[25] To which Aiken replied that he had his orders. "Well, I'll get that corrected immediately." Within "just a matter of hours" he "had orders to pack up and go back to Harvard and become the officer in charge of the United States Naval Computing Project. I guess I'm the only man in the world who was ever Commanding Officer of a computer."

But let us not denigrate this amazing piece of equipment. Standing eight feet high, and stretching 51 feet, it was a sight to behold: starting from the left, your eyes are struck by two giant banks of dials. The first bank sets the constant registers, the second the storage registers. Then endless columns of plug-in relays, each of which must be set by hand, until finally a selection of IBM telewriters type out the results. And that's from the front: looked at from behind, all that could be seen were forests of interconnecting wires. Little wonder that it took the five-strong team of IBM engineers several weeks to reassemble it.

The machine found a home in the basement of Harvard's Research Laboratory of Physics, displacing a gigantic wet-cell battery (much to the annoyance of some in the lab). Under Campbell's supervision, and with the help of Robert Hawkins and researcher David Wheatland, the calculating machine had already been put to work. For instance, Aiken's colleague Ronald King asked it to calculate integrals.

As with all early computers, however, the Mark I didn't always behave well. Hawkins described himself as being on call "24 hours a day, seven days a week"[26] to tackle problems, and described one occasion when it was offline for three days. "I stayed there – I don't know how long – but I just couldn't keep my eyes open anymore. I just told the boss, 'I'm going home, I don't care!'" On his return, he and the rest of the team "fixed it right away".

---

[24] I Bernard Cohen, *Howard Aiken: Portrait of a Computer Pioneer*, p160
[25] Howard Aiken oral history interview with Henry Tropp and I B Cohen, 26 February 1973, Smithsonian National Museum of American History, **rpimag.co/aikeninterview**, p54
[26] Robert Hawkins oral history interview with William Aspray, 20 February 1984, Charles Babbage Institute, **rpimag.co/hawkinsinterview**, p13

IBM's Durfee stayed with the machine in those early days. Days during which they still didn't trust the machine, having grown used to checking its results using desk calculators. It was a tough habit to break, but eventually the team grew to rely on the Mark I's built-in checking mechanisms.

In contrast to the Colossus or ENIAC, there was no veil of secrecy covering the Mark I. IBM was keen to promote the role that it had to play in the creation of this war machine, and so it's easy to see why Watson was so furious when he read Harvard University's press release about the project. One that the US Navy had approved, but which Harvard failed to run past IBM.

Under the heading "World's greatest mathematical calculator", its first paragraph read in uncontroversial fashion: "The world's greatest mathematical calculating machine, a revolutionary new electrical device of major importance to the war effort, will be presented today to Harvard University by the International Business Machines Corporation to be used by the Navy for the duration."

Nor could IBM have found much to fault in the following three paragraphs, which described the forthcoming ceremony and the machine itself. For example: "An algebraic super-brain employing a unique automatic sequence control, it will solve practically any known problem in applied mathematics. When a problem is presented to the sequence control in coded tape form it will carry out solutions accurate to 23 significant figures, consulting logarithmic and other functional tables, lying in the machine or coded on tapes."

The problem this computer couldn't solve came in the fifth paragraph, which described Aiken as "the inventor… who worked out the theory which made the machine possible." And while the release went on to namecheck both Frank Hamilton and Benjamin Durfee, and stated that the construction "work was carried on at the Engineering Laboratory of the International Business Machines Corporation at Endicott, New York, under the joint direction of Commander Aiken and Clair D Lake" this reportedly sent Watson into a fury when he read it.

The controversy stems from two interweaving issues. First, IBM hotly disputed the seemingly simple statement that Aiken was the inventor of the Mark I. While it's true that the machine wouldn't have existed without Aiken, an uncountable number of design decisions were taken in partnership with IBM. Many without Aiken, while he was seconded to the Navy.

Second, where was the credit for IBM the company? Yes, it and three members of its staff were mentioned, but only fleetingly, and the release made it sound like IBM passively built the machine to Aiken's specification. There was no sense of a six-year partnership, the intellectual contribution from Bryce and others, the scale of IBM's financial investment, nor the key role played by benefactor Watson – without whose sign-off the machine would never have been built.

Grace Hopper
Image: US Department of Defense, Public Domain

It didn't help that Harvard failed to even meet Watson and his wife when he emerged from the train in Boston. He had expected to be met by "some Harvard dignitaries, or their emissaries, with an official limousine"[27] but was instead greeted by IBM's Boston branch manager, Frank McCabe, in his two-door Chevrolet. McCabe later recounted the difficulty Watson's wife had getting into the car's back seat, but this was nothing compared to the anger McCabe saw growing on Watson's face as he read the press release.

We can only imagine Watson's mood when he summoned Aiken and Conant to his hotel room that evening, and the invective he used to berate Aiken in particular. While there is some speculation that some of this show of anger may have been calculated – Watson once stated that he used such tactics to get results in his business dealings – what we know for sure is that it had immediate effect. Harvard quickly reworded and reissued the press release, and Aiken's address at the official unveiling the following day was fulsome in its praise for IBM, Watson, and the company's engineers.

The short-term effect of this fallout was limited – Watson even went ahead and gave his generous $100,000 gift towards the ongoing running costs of the machine – but this would be the end of the partnership. IBM would go on to invent its own

---

[27] I Bernard Cohen, *Howard Aiken: Portrait of a Computer Pioneer*, p123

super-calculator, the Selective Sequence Electronic Calculator, determined that it was superior in every way to the Mark I. And it's doubtful whether Aiken, a proud man, would have wanted anything more to do with IBM after that point either.

Nor would the following day's newspaper coverage have helped. While many reports mentioned that the computer was given to Harvard by IBM, others said Aiken handed the computer to Harvard and many barely mentioned the company's role, instead focusing on a narrative that the computer's existence was due to Aiken being too lazy to work out the results of his thesis by hand – based on an interview Aiken gave to reporters prior to the official presentation.

At least the photographs of the finished computer, usually showing Navy ensigns at work on it, included the stylish front design commissioned by Watson that gave this machine such a striking look. Designed by industrial designer Norman Bel Geddes, famous for futuristic concepts such as teardrop-shaped vehicles, Grace Hopper said it "gave poor Howard Aiken an awful pain, because it was fifty or a hundred thousand bucks for the case, and he could've used it for a computer, and that irked him."[28]

A lieutenant (junior grade) in the Navy Reserves, Hopper joined the project on 2 July 1944, adding her intellectual weight to the Harvard operatives – including Robert Campbell, Robert Hawkins, and Physics Research Associate David Wheatland – and Richard Bloch, an April addition who reported directly to a Rear Admiral. Of Bloch, Hopper later said that he was "the only person I ever knew who wrote a program in ink and was correct the first time. But Bloch just thought like the machine did."[29][30]

It was Bloch who was put in charge of the Mark I's most famous series of calculations, not that he was told why at the time. All he knew, other than the technical description of the nonlinear differential equation in question, was that the problem described "a spherically symmetric flow of a compressible fluid in the presence of a spherical detonation wave proceeding inward from the surface of the sphere towards its centre".[31] Although clues in its importance came via the presence of John von Neumann, then

---

[28] Grace Hopper oral history with Uta Merzbach, 7 January 1969, Smithsonian National Museum of American History, **rpimag.co/hopperinterview1969**, p8

[29] Grace Hopper oral history with Angeline Pantages, December 1980, Computer History Museum, **rpimag.co/hopperinterview1980**, p23

[30] In his entertaining yet boringly titled essay, 'Reminiscences of Aiken during World War II and Later', Bloch disputes his infallibility, describing Grace Hopper's description as "something of an exaggeration". See p197 of I Bernard Cohen and Gregory W Welch, *Makin' Numbers: Howard Aiken and the Computer* (The MIT Press, 1999, ISBN 978-0262032636)

[31] I Bernard Cohen, *Howard Aiken: Portrait of a Computer Pioneer*, p164

Card punch used to prepare programs
Image: Arnold Reinhold, CC BY-SA 3.0

a key physicist in the Manhattan Project, whenever the Mark I was working on the problem. He only discovered the exact reason for the secrecy when news of the first atomic bomb, which laid waste to Hiroshima, became headline news in 1946.

This doesn't mean the bomb only happened because of the Mark I. In their 1982 essay, 'Early Computing at Los Alamos,'[32] Nicholas Metropolis and Eldred Nelson revealed that Los Alamos physicists solved the problem in three weeks using a series of existing IBM equipment, while the Mark I was "halfway through the problem" after five weeks. It wasn't a fair race – the Los Alamos team produced answers to six digits while the Mark I was accurate to 18 – but it does highlight that even by 1945 standards the computer was relatively slow.

Partly, the slow speed was by design. Aiken decided at an early stage that the computer would calculate to 23 digits, with the 24th place indicating whether the number was positive or negative. But that degree of accuracy was rarely, if ever, required, meaning the Mark I laboured through its calculations for significantly longer than was necessary. Another key reason was the computer's reliance on electromechanical relays rather than the electronic valves inside the ENIAC and Colossus, which inevitably adds

---

[32] Nicholas Metropolis and Eldred Nelson, 'Early Computing at Los Alamos', October 1982, in the *IEEE Annals of the History of Computing*, Vol 4 Issue 4, pp348-357

Program tape with visible programming patches
Image: Arnold Reinhold, CC BY-SA 3.0

time. It also adds noise, with a Harvard student of that time, Jeremy Bernstein, eloquently, albeit with a tinge of sexism, describing the Mark I as sounding "like a roomful of ladies knitting" when in operation.[33]

Some would argue that this means the Mark I wasn't really a computer, just a supercharged, interlinked version of the electromechanical calculators that had gone before. Critics might also correctly point out that it didn't store programs: the sequence of operations depended on a program punched out on tape.

Where it wins out over some more illustrious rivals, however, is that it was completed and put to use before World War II was finished (even though the war was not on Aiken's mind in 1937 when he first conceived the idea). When Aiken gave a talk in 1960 covering its wartime tasks,[34] he mentioned "calculations in the development of radar", helping to "solve blast furnace difficulties" and calculating tables "of tremendous importance in magnetic mine warfare". But he didn't mention the Los Alamos work, so may have kept back some still-classified projects.

It was certainly put to heavy use, with three eight-hour shifts filling its day when it had an active problem to work through. And after Japan surrendered to the Allies on 14 August 1945, the Mark I moved on to its second phase in life: as a scientific, academic computer. And like so many early computers, it would evolve. The Mark I would eventually benefit from 24 more storage registers (taking it to 96), the addition of a high-speed multiplying unit, the ability to do calculations at double precision (46 digits) or half precision (11 digits) to save time. And, arguably most crucially, the introduction of program branching.

---

[33] Jeremy Bernstein, *The Analytical Engine* (Secker & Warburg, 1965), p66
[34] I Bernard Cohen, *Howard Aiken: Portrait of a Computer Pioneer*, p167

All these enhancements, along with greater reliability when every single one of its 3,500 relays were replaced, meant it continued to be put to good use until it was finally decommissioned in 1959.

Photos of the Harvard Comp Lab in the 1950s show it sitting alongside its successors, which were created for military use rather than to replace the Mark I. Instead of heading down the fully electronic route with the Mark II, which was built between 1945 and 1947 for the US Navy, Aiken and his team decided to effectively create a bigger, better version of the Mark I. It would consume twice as much floor area – 4000 square feet – and could operate on two problems at once. In his essay, 'Mark II, an Improved Mark I,'[35] Robert Campbell refers to a left side and a right side, that could combine forces on a larger problem if necessary. Effectively, it was two computers in one.

Instead of calculating to 23 digits, the Mark II calculated to ten digits and supported floating points. This, Aiken felt, was better suited to its primary role of calculating firing tables. They also built it at Harvard, working with third-party companies to manufacture specialist components such as relays. These remained the biggest handbrake in terms of calculation speed, but the Mark II was still six times faster at calculations than the Mark I, according to Campbell.[36]

Mark III and IV were to follow, the III designed for the Navy and the IV for the US Air Force, but that was the end of the line for Aiken's computers. By the late 1950s, the EDVAC-based design – or von Neumann architecture, to put it another way – had taken hold. There is no place for the Mark I on a modern-day computer's family tree.

Still, Aiken's creation continues to have a ripple effect to this day.

Most of this effect stems from his decision to create a Computer Lab (almost immediately shortened to Comp Lab in the Harvard vernacular) and a formal Computer Science education for Harvard's students, which in turn created a rich feeding ground for IBM and its rivals over the years. Many of Aiken's acolytes would go on to successful careers at the company, with Thomas Watson Jr proving keen to restore goodwill between IBM and the university soon after he succeeded his father as chairman in 1952.

---

[35] Robert Campbell, 'Mark II, an Improved Mark I', in I Bernard Cohen and Gregory W Welch, *Makin' Numbers: Howard Aiken and the Computer*, pp111-127
[36] As above, p122

Then there is the Harvard computers' impact on programming as a whole. Much of the credit here goes to Grace Hopper, who wrote the Mark I's manual – this necessarily included detailed instructions on how to program it. It was a task fraught with difficulty, with Hopper famously labelling the moth they found trapped in one of the Mark II's relays as "first actual case of bug being found".

That happened on 9 September 1947 (a fact we happen to know because Hopper taped the moth into the logbook), but as Peggy Kidwell explained in her 1998 article, 'Stalking the Elusive Computer Bug',[37] the term 'bug' first emerged in the late 19th century. Thomas Edison even used the term in relation to the "little faults and difficulties" that must be ironed out in the process of taking an invention to commercial reality.

Alas, we can't precisely give the term 'debug' a vigorous etymology, but once again Grace Hopper must take some credit for its popularisation. She left Harvard in 1949 to join the Eckert-Mauchly Computer Corporation, and when writing a glossary for the UNIVAC she defined the term as a colloquialism "to remove a malfunction from a computer or an error from a routine."[38] Such as, one might suppose, a squashed moth from within a mechanical relay.

---

[37] Peggy Aldrich Kidwell, 'Stalking the Elusive Computer Bug', in *IEEE Annals of the History of Computing*, Vol 20, No 4, 1998, p5
[38] Grace Hopper, 'A Glossary of Computer Terminology', in *Computers and Automation*, Vol 3, No 5, May 1954, p16

ENIAC computer on display
at Fort Sill Museum, Lawton,
Oklahoma, USA

Image: Judson McCranie,
CC BY-SA 3.0

ENIAC
(Electronic Numerical Integrator And Computer)

**Built to do military calculations,
it was the first general-purpose
digital computer**

If war stymied the development of the ABC, it was the making of the ENIAC. Its story can be simplified to one sentence: the US Army needed a more efficient way to calculate firing tables for its weapons, conventional methods couldn't keep up, and a bright young professor called John Mauchly had the answer in the form of an electronic computer.

The real story is far more complicated. One that has also been camouflaged by lawsuits, huge corporations, warring egos, and the breadth of time. Here, we will attempt to navigate our way to the truth – but, in the same way that 1940s firing tables could only offer accuracy within certain bounds, we must do the same.

A function table from ENIAC, on display at the US Army Ordnance Museum
Image: Judson McCranie, CC BY-SA 3.0

One thing we can state with confidence is that the ENIAC would have remained a concept were it not for John Mauchly meeting Herman Goldstine, a mathematician and officer at the US Army's Ballistic Research Laboratory (BRL). Goldstine saw his primary duty as "finding new apparatus to expedite the production of firing and bombing tables,"[1] which were in hot demand from the US Army and Air Force.

We will let Harry Polachek, who headed the BRL group that created the Army's firing tables, set out the scale of the challenge.[2] Using a desk calculator, he explained, "it required an average of two eight-hour days to compute the path of a single trajectory. To compute a firing table, it was necessary to find the solution to hundreds of trajectories in addition to carrying out extensive auxiliary computations of equal difficulty." Depending on the weapon in question, that could take up to three months.

When Polachek joined the BRL in March 1941, the lab had a dozen staff. That itself was double the number from a year earlier. The US had not yet joined the

---

[1] Herman H Goldstine, *The Computer: from Pascal to von Neumann* (Princeton University Press, 1993, ISBN 978-0691023762), p138
[2] Harry Polachek, 'Before the ENIAC [weapons firing table calculations],' in *IEEE Annals of the History of Computing*, Vol 19, No 2, Apr–Jun 1997, doi: 10.1109/85.586069, pp25-30

conflict – that would come four days after Japanese planes attacked Pearl Harbor on 7 December – but it was already obvious to the US military that war was only a matter of time. They needed new firing and bombing tables to accompany their new weapons.

A year earlier, the director of the BRL – Colonel Herman H Zornig – had visited the University of Pennsylvania, in particular the Moore School of Electrical Engineering, to request use of its differential analyser to help create firing tables. While differential analysers weren't as accurate as desktop calculators, they could work a single trajectory much more quickly: "about 15 to 30 minutes" according to Polachek, although he added the caveat that "it required a lengthy period (a day or more) to change from one type of trajectory to another."

Between them, the BRL and Moore School owned two out of three differential analysers in existence in the US at the time, but it wasn't enough. More manpower was required; or to be accurate, more women power. Men were already being drafted into the Army, and so when the BRL asked the Moore School to provide "a staff of human computers"[3] it was happy to cooperate. "Moore School immediately agreed, and the staff, primarily college graduates who could handle some mathematics, quickly grew to more than 100," said John Grist Brainerd, a professor at the Moore School who was in charge of liaison with the BRL.

However, the computers' work was dull. "It's like doing your income tax every day of your life, from morning till night," Goldstine explained in an interview with historian Nancy Stern.[4] While the Army wanted the women to be based in Aberdeen, the women themselves were less keen. "Aberdeen was about 32,000 acres of swamp, and the town itself was a little wide place in the road… and it was pretty much a hellhole," said Jean Bartik, one of the ENIAC's initial programmers.[5] Why would someone want to stay there, in barracks, when they could live in an apartment in central Philadelphia?

The answer was they wouldn't, instead working out of the Moore School, and this turned out to be fortuitous. On one of his regular trips to check on the computers, Goldstine spoke to John Mauchly's ex-student Joseph Chapline – who would go on

---

[3] John G Brainerd, 'Genesis of the ENIAC,' in *Technology and Culture*, Vol 17, No 3, July 1976, pp482-488
[4] Interview with Herman Goldstine by Nancy Stern on 14 March 1977, Niels Bohr Library & Archives, American Institute of Physics, College Park, MD USA, **rpimag.co/goldstineinterview**
[5] Interview with Jean Bartik by Gardner Hendrie on 1 July 2008, Computer History Museum, reference number X4596.2008, **rpimag.co/bartikinterview**

Cpl. Irwin Goldstein sets the switches on one of ENIAC's function tables at the Moore School of Electrical Engineering
US Army photo, 1946, Public Domain

to write an operating manual for the BINAC that is said to have set the template for computer manuals thereafter. Chapline suggested that Goldstine seek out his former lecturer, who had long been espousing the creation of electronic computers to crunch numbers, and wanted to create a machine capable of forecasting the weather.

Mauchly recalled his and Goldstine's first meeting and discussion about an electronic computer in a separate interview with Stern.[6] "After we'd talked a little bit, he [Goldstine] said, 'You ought to write this up.' I said, 'I have.' 'Well, where is it?' 'Well, I don't know. I'll look.'"

Mauchly had sent the memo to various members of the faculty, including Brainerd, but their reaction had been lukewarm at best. By the time Mauchly met Goldstine for the first time, not only had Mauchly stopped pushing the idea (in writing at least) but all copies of the memo had been lost.

---

[6] Interview with John Mauchly by Nancy Stern on 6 May 1977, Niels Bohr Library & Archives, American Institute of Physics, College Park, MD USA, **rpimag.co/mauchlyinterview**

This may seem odd in the light of the ENIAC's importance. Surely Mauchly would be feverishly attempting to turn his electronic computing concept into reality? After all, he was passionate enough about it to drive a thousand miles to visit Dr Atanasoff, as we heard in the story of the ABC, on discovering the Iowa State professor had created a prototype digital computer.

However, according to Goldstine, this fits Mauchly's character as an ideas man rather than a finisher. "Wilks [Samuel Wilks, an American mathematician] said Mauchly would occasionally drive down… and spend the day talking to Wilks about his ideas on applying methods of statistics and probability theory to meteorology, and then would just disappear, and nothing would come out of it in the form of a paper. This is just not… the normal academic method."[7]

Goldstine added: "I'm not saying this in any disrespect of Mauchly's abilities. I'm merely commenting that he was not a person who was good at starting something and seeing it through to fruition… I think that's why Mauchly could have written many proposals to no effect, because he always did them to perhaps entertain himself."

It certainly explains how six months later his proposal for an electronic computing device had evaporated into thin air. Fortunately, Mauchly's secretary still had the shorthand notes for the original memo, and she quickly reconstructed the document. Over the course of the next few weeks, with the help of an outstanding young engineer called J. Presper Eckert, that memo turned into a formal proposal for the building of the world's first general-purpose electronic computer.

On 9 April 1943, Eckert's 24th birthday, Goldstine and Brainerd took the proposal to the Ballistic Research Laboratory at Aberdeen. Despite the proposed $150,000 budget, Colonel Paul Gillon, who was responsible for the computing branch of the BRL, was quick to give the project his backing.

"He [Gillon] was convinced that the computing load at the Proving Ground was not going to be eased by any change of factors of two or three, or things like that,"[8] said Goldstine. "He saw that what really was needed was something like an order of magnitude, and he was prepared to go forward with almost anything, if it was only reasonable."

---

[7] Interview with Herman Goldstine by Nancy Stern on 14 March 1977, Niels Bohr Library & Archives, American Institute of Physics, College Park, MD USA, **rpimag.co/goldstineinterview**

[8] As above

It would be wrong to give Colonel Gillon all the credit. Oswald Veblen, then the chief scientist at the BRL and a distinguished Maths professor at Princeton University, was another big name to back the project. In his book, *The Computer: from Pascal to von Neumann*,[9] Goldstine describes a BRL meeting in the spring of 1943 involving him, Veblen, and Colonel Leslie Simon, where "Veblen, after listening for a short while to my presentation and teetering on the back legs of his chair brought the chair down with a crash, arose, and said, 'Simon, give Goldstine the money.'"

Still, others in the Army were harder to win around. It would require a huge number of vacuum tubes, still in short supply, and some senior figures believed two alternative projects would offer results more quickly. First, Vannevar Bush's Rockefeller Differential Analyzer, an electronic version of the differential analyser that required minutes rather than a full day to set up for a different weapon. Second, the Harvard Mark I, an electromechanical computer built upon the principles set out by Charles Babbage's analytical engine a century earlier.

Arguably, the doubters turned out to be correct. An early version of the Rockefeller Differential Analyzer was already in service by early 1943, and it was said to be one of America's secret weapons for the war, such was its effectiveness. The Mark I proved less impactful during the war, but we cover this remarkable IBM-backed computer in more detail in Chapter 5.

Despite the reluctance in some quarters, Gillon successfully fought for the electronic computer project, securing a budget of $61,700 for six months. This initial sum would cover "research and development of an electronic numerical integrator and computer and delivery of a report thereon," later wrote Martin H Weik. It was this document, and Gillon, that gave the ENIAC its name.

The agreement for 'Project PX' was signed on 5 June 1943, eight weeks after the official pitch meeting with Colonel Gillon, but this was a formality. A week earlier, Goldstine, together with Professor Brainerd, had formed a team. At its head stood Eckert, the chief engineer, with Mauchly the principal consultant (he would continue in his teaching role).

---

[9] Herman H Goldstine, *The Computer: from Pascal to von Neumann*, p149

Detail of the rear of a section of ENIAC, showing vacuum tubes
Image: Paul W Shaffer, CC BY-SA 3.0

They were initially joined by a handful of others, including Arthur Burks, Thomas Kite Sharpless, and Jack Davies. Under Eckert's direct and demanding supervision, each was given a section of ENIAC to design.

At this point, the ENIAC was a concept rather than a blueprint. The team didn't even know what it would look like. But they had already settled on the three key areas: one for mathematical operations, one for storing data, one for programming.

Now, the small team needed to turn theory into practice. That meant devising and testing designs (Burks said the team initially worked on "basic counting circuits and switching circuits – building different counters and testing them"[10]), and finding out existing resources they could draw upon and what they needed to invent themselves. For example, Goldstine describes visits to the RCA Research Laboratories in Princeton, New Jersey, where they saw a function table ("a way of storing a table of

---

[10] Tom Infield, 'Faster than a speeding bullet', in *The Philadelphia Inquirer*, 4 February 1996, p26

Chapter 6: ENIAC   115

numbers in the form of an electrical network of resistances"[11]) that would eventually become a key part of the ENIAC.

Mauchly and Eckert had long been aware of 'flip-flop' circuits, where a pair of vacuum tubes worked as an electronic on-off switch to represent zero or one. This idea dated back to 1918, when two British physicists – William Henry Eccles and Frank Wilfred Jordan – filed a patent for what would become the Eccles-Jordan trigger circuits.

The bigger issue was reliability. Vacuum tubes were already commonplace in radios, but they were prone to failure because they were constantly heated and cooled, much like domestic light bulbs. Even at this early stage, Eckert and Mauchly knew their computer would require thousands of vacuum tubes, and the ENIAC would be rendered useless if tubes had to be constantly replaced.

Indeed, one of Burks's first jobs was to see if the ENIAC could use thyratron tubes – filled with gas – rather than vacuum tubes, to see if they would be reliable at high speeds. While he eventually succeeded at making the circuits work at 100,000 pulses per second, the ambitious target set by Eckert, they simply weren't reliable enough[12] and Burks's work was abandoned. Another reason the team settled on vacuum tubes was they discovered they "would last a lot longer if we kept them below their proper voltage – not too high or too low," according to Eckert in 1989.[13]

Eckert credits the fact that they started slowly as one of the key reasons for the ENIAC's success, with the team laying the foundations in steps rather than rushing to build anything approaching a full system. They only started building the first two accumulators several months into the project, having secured a second round of funding. Eventually, ENIAC would include 20 such accumulators working in tandem, bringing to life one of Mauchly's core visions: the power of 20 desk calculators working in parallel.

Desk calculators, explained Goldstine, "contained [mechanical] counter wheels which could be turned one stage at a time by the reception of an electric signal. Electronic ring counters are analogous to these wheels." They were called ring

---

[11] Herman H Goldstine, *The Computer: from Pascal to von Neumann*, p163
[12] Interview with Alice R Burks and Arthur W Burks by Nancy Stern on 20 June 1980, Charles Babbage Institute. Retrieved from the University of Minnesota Digital Conservancy, **hdl.handle.net/11299/107206**
[13] 'From ENIAC to Everyone: Talking with J Presper Eckert', Alexander Randall, 23 February 2006, **rpimag.co/fromeniac**

counters because the final output fed into the first input of the next digit, akin to carrying over a number in mental arithmetic.

When a new number was fed into an accumulator – via a train of pulses – it was then added to (or subtracted from) the existing number. Each accumulator included ten ten-digit counters – called decade counters – so could hold a number up to 10^{10}, otherwise known as 10 billion. To be precise, 9,999,999. Along with addition and subtraction, the ENIAC included dedicated units for high-speed multiplication, division, and square rooting.

The ENIAC could compute an addition or subtraction in 200 microseconds, so 1/5000th of a second. Division and square-rooting were more complicated, taking around 3/100ths of a second, while multiplication took up to three milliseconds. By contrast, a desk calculator could complete a multiplication in ten seconds at best. Even the Mark I took around three seconds for a ten-digit complication.[14] In comparison to other methods of the time, the ENIAC was a mathematical beast.

The team completed the first two accumulators in mid-1944, at which point they felt confident that not only would they complete the project but they were on the verge of something special. "It soon emerged," wrote Goldstine, "that the machine would be much more useful than just a device for solving the differential equations of exterior ballistics. It gradually became clearer that the great advantage of the digital approach was that the ENIAC was going to be a truly general-purpose device."[15]

Even with just a pair of accumulators up and running, they could perform arithmetical problems and solve basic differential equations (the key to calculating firing tables). Meanwhile Jeffrey Chuan Chu, another engineering graduate drafted into the growing ENIAC team from the Moore School, set to work on the divider and square-rooter.

However, the team still needed help for what we would now call input/output, or I/O. And that meant working with IBM to devise a way to integrate its industry-standard punch cards. Colonel Gillon visited IBM's chairman Thomas Watson Sr in February 1944, who helped to arrange a meeting between his engineers and those at the Moore School that eventually led to a solution – the constant transmitter.

---

[14] Herman H Goldstine, *The Computer: from Pascal to von Neumann*, p137
[15] As above, p163

"The constant transmitter with its associated card reader reads from punched cards, numbers that are changed in the course of a computation and makes these numbers available to the computer as needed," explained Adele Goldstine in her comprehensive 'Report on the ENIAC'[16] for the US Army's Ordnance Department, printed in 1946.

One apparent backwards step compared to the binary ABC was for ENIAC to use base ten (decimal), but Eckert was quick to explain why in a 1977 interview.[17] "There were many sub-elements of the original ENIAC that are binary in nature and then could be converted to decimal before they came out of the machine. It was a conscious decision forced on us by the fact that the IBM card punch and printers and things that we hooked to were decimal. It had nothing to do with anything but that."

If you can detect a certain spikiness in Eckert's reply, your instincts are correct. Moments earlier in that interview he said: "Some people seem to think that... we didn't learn about the binary system. I assure you that I was familiar with the binary system when I was 14 years old."

While Mauchly came up with many of the ideas, and was a key decision-maker throughout the project, Eckert oversaw day-to-day operations for good reason. He was a stickler for getting things right, overseeing the project with exceptional levels of detail.

"I took every engineer's work and checked every calculation of every resistor in the machine to make sure that it was done correctly," he told historian Nancy Stern.[18] "Normally, I wouldn't want to have to do that. But this was the first time for a machine with in the order of 100 times as many tubes as anybody has ever built electronically. And if it was going to work, one had to be 100 times more careful."

No detail was too small. In an interview with Alexander Randall,[19] Eckert describes how they protected themselves from a basement's worst enemy: mice. "We knew mice would eat the insulation off the wires, so we got samples of all the wires that

---

[16] Adele K Goldstine, 'A report on the ENIAC', 1946, pp1-12. Retrieved from Smithsonian Libraries, **library.si.edu/digital-library/book/reportoneniace100moorb**

[17] Interview with John Presper Eckert by Nancy Stern on 28 October 1977, Charles Babbage Institute. Retrieved from the University of Minnesota Digital Conservancy, **hdl.handle.net/11299/107725**

[18] Interview with John Mauchly by Nancy Stern on 6 May 1977, Niels Bohr Library & Archives, American Institute of Physics, Maryland, **rpimag.co/mauchlyinterview**

[19] Alexander Randall, 'From ENIAC to Everyone: Talking with J Presper Eckert', 23 February 2006, **rpimag.co/fromeniac**

Two pieces of ENIAC on display in the Moore School of Engineering and Applied Science
Image: Paul W Shaffer, CC BY-SA 3.0

were available and put them in a cage with a bunch of mice to see which insulation they did not like. We only used wire that passed the mouse test."

Carl Chambers, a research director at the University of Pennsylvania, paints a vivid picture of Eckert as supervisor.[20] "There wasn't a single one of the staff who was doing a breadboard setup that he didn't tell him where to solder the joint," he said. "And he'd come in the next morning and tell them to tear everything up and change it because it was going to be revised as a result of his overnight idea for improvement."

This might normally cause friction within a team, but Chambers claims that "Eckert's knowledge of circuits and so on was superior. They were glad to get the advice, glad to get the ideas, glad to learn from him."

Despite the workload, group spirit remained strong throughout the project. "Everybody felt this [building the ENIAC] was terribly important to the war and

---

[20] Interview with Carl Chambers by Nancy Stern on 30 November 1977, Charles Babbage Institute, University of Minnesota, Minneapolis. Retrieved from the University of Minnesota Digital Conservancy, hdl.handle.net/11299/107216

that we get it done," said Arthur Burks,[21] one of the first members on the team. "So nobody hesitated to work all day and then all evening or come in on the weekends if necessary. Typically we worked six days a week and didn't come in on Sundays, I recall, until things later became very busy and then we would come in on Sunday."

Unfortunately, the team's hopes that they would finish work at some point in 1944 proved impossible. And like so many information technology projects, as time wore on so costs increased. The initial estimate was $150,000, but the final cost came in at $487,000.

"Part of this growth was due to expansions of requirements," explained Mauchly at the Los Alamos International Research Conference in 1976.[22] "[As] others saw what we were doing and how we were going about it, they said, 'We don't want just one of these function tables to put in drag functions, maybe for that really flexible, general-purpose device we ought to have three.' So they wanted more accumulators, too, more ten-digit storage devices."

The growth Mauchly refers to is literal too. The plan had always been for the ENIAC to be located in the Aberdeen Proving Ground, and Goldstine first asked for a room "about 20 feet by 40 feet to house the machine." This proved too small, with the computer eventually consuming the 30-by-50 foot (139m^2) basement of the Moore School. The BRL ended up building a dedicated annex to house the world's first computer.

All this work happened under the military cloak of secrecy, with even the female computers left to wonder what was being built. The only clues, explained Kay Mauchly Antonelli,[23] was a trail of workmen carrying metal panes into a room. Until one evening, when she was working a night shift on the analyser. "Eckert and Mauchly came down to the analyser room and said, 'Would you like to see what we've been working on?' They said they had achieved a milestone."

The milestone was hooking up the two accumulators so they could communicate, with the men demonstrating this with a simple calculation of five times one thousand. "They were very elated," said Antonelli. She, however, was rather less impressed. "It

---

[21] Interview with Alice R Burks and Arthur W Burks by Nancy Stern on 20 June 1980, Charles Babbage Institute. Retrieved from the University of Minnesota Digital Conservancy, **hdl.handle.net/11299/107206**

[22] Video recording of 'The ENIAC by John W Mauchly', The International Research Conference in the History of Computing at Los Alamos, 1976, **youtu.be/OUsc5JnyBYU**

[23] Tom Infield, 'Faster than a speeding bullet', in *The Philadelphia Inquirer*, 4 February 1996, p26

Glenn Beck and Betty Snyder program the ENIAC in building 328 at the Ballistic Research Laboratory
US Army photo, c. 1947–1955, Public Domain

didn't mean anything to me. I had no concept of how this would fit into the running of a trajectory."

By June 1945, it was time to start training the programmers. That is, the people who would interpret the equations and plug in the appropriate cables. The team started small: Kay McNulty (as she was then called; she would marry John Mauchly in 1948), Betty Holberton (née Snyder), Marilyn Meltzer (née Wescoff), Ruth Teitelbaum (née Lichterman), Frances Spence (née Bilas) and Jean Bartik (née Jennings). All six were sent to Aberdeen for a two-month training session.

It was an inauspicious start. "They just gave us these block diagrams of the ENIAC and told us to study them, learn how it works."[24] The women weren't allowed to see the ENIAC itself as their security clearance was too low. Bartik and Betty

---

[24] Interview with Jean Bartik by Gardner Hendrie on 1 July 2008, Computer History Museum, p20. Retrieved from **rpimag.co/bartikinterview**

Programmers Jean Bartik (left) and Frances Spence operating ENIAC's main control panel
US Army photo, c. 1945, Public Domain

Holberton picked one of the empty classrooms and "just sat there with this block diagram and we didn't even know how to read it."

Fortunately, a man soon walked in and introduced himself as John Mauchly. "So we almost fell off our chairs," said Bartik, as both women were well aware of who he was. "And we said, 'Boy, are we glad to see you?' You know, 'tell us how this blasted accumulator works'. Well, anyway, he was a marvellous teacher, absolutely a wonderful teacher."

He needed to be, because programming the ENIAC was an incredibly complex job. Especially when it became clear that the women would need to not only understand the maths – something they were more than qualified for as college graduates in the topic – but also translate that into settings and then the physical work of plugging everything in. This was an arduous, complex task that could take hours.

And time was ticking. Despite everyone's hard work, the ENIAC project stretched beyond World War II, with Victory over Japan Day marking its official end for the

USA[25] on 2 September 1945. So rather than calculating ballistics trajectories, ENIAC's first proper test started a month later with a problem set by scientists from Los Alamos, which was now working on the hydrogen bomb. "Our problem was with one dimension in space and one dimension in time, of course, to study some of the thermonuclear possibilities," said Nicholas Metropolis,[26] an experimental physicist who played a pivotal role in the Manhattan Project. "These were the first realistic, semi-realistic tests."

Famously, John von Neumann was also part of the Manhattan Project and had, by chance, become familiar with the ENIAC after Goldstine had spotted him at Aberdeen train station. While von Neumann played no notable part in the ENIAC's development, he had met both Mauchly and Eckert, was familiar with the computer's workings and would play a big part in the creation of the EDVAC. He and Edward Teller, the so-called 'father of the hydrogen bomb', suggested that they run a model on the ENIAC, although the actual setup fell to Metropolis and computer scientist Stanley Frankel.

The pair had learned enough about ENIAC to program the problem themselves, but the physical plugging in of cables and switch setting was done by the team of programmers. And while the calculations never took long due to ENIAC's power, the entire process took weeks, with Metropolis and Frankel starting work in late autumn 1945 and not leaving until January the following year. In April, the results were reviewed by a "blue-ribbon panel of current and former Los Alamos scientists,"[27] producing a final report that said a hydrogen bomb "would probably work."

It's testament to everyone behind the ENIAC project that a computer commissioned to create firing tables could tackle such a radically different problem. Right from the start, though, the team were thinking long-term, with many of their implementations echoing down to modern computing. "[Mauchly] pointed out that these problems we were doing were highly repetitive and that obviously you couldn't store the program by cables or switches," said Eckert. "However, we planned to do [solve the problems] over and over again. But what you should do is store the program once and then call on it and use it over."

Programmers will recognise these repeated tasks as subroutines, with Eckert giving Mauchly the credit for inventing them – if not the term. "The ENIAC was

---

[25] In the UK, Victory over Japan (VJ) Day is commemorated on 15 August.
[26] Interview with Nicholas Metropolis by William Aspray on 29 May 1987, Charles Babbage Institute. Retrieved from the University of Minnesota Digital Conservancy, **hdl.handle.net/11299/107493**
[27] Steve Leibson, 'Stanley P Frankel, Unrecognized Genius', **hp9825.com/html/stan_frankel.html**

nicely equipped with the ability to handle subroutines," said Mauchly at the Los Alamos conference, quoting a paper written by Eckert (who refused to turn up as he was to be introduced by John Grist Brainerd, a man he had come to detest). "I think we should be particularly proud that we did have this flexible programming and we did not hamper the people who wanted to put programs on."

Jean Bartik believes some of the credit for subroutines should go to Kay McNulty. The two were "trying to figure out how the ENIAC could do a trajectory" when they realised there wasn't enough hardware. "And finally I remember one day Kay said, 'Oh, I know, I know, I know. We can use a master programmer to repeat code,' and we did. We began to think about, you know, how we could have subroutines, and nested subroutines, and all that stuff."

In February 1946, the US Army was ready to show the ENIAC – until then a confidential project – to the world. "I was in charge of the demonstration," said Arthur Burks in a 1974 talk.[28] "Seems a bit silly, but I told the press, 'I am now going to add 5000 numbers together' and pushed the button. The ENIAC added 5000 numbers together in one second. The problem was finished before most of the reporters looked up!"

The problem chosen for a demonstration two weeks later, to a roomful of military VIPs, was rather more challenging: proof that the ENIAC could indeed calculate a trajectory of a missile. When Goldstine asked Betty Holberton and Jean Bartik at the start of February if the calculation was ready, so that it could be used in the demonstration, Bartik says they replied "You bet."[29] Except it wasn't. "So we went back and began working like mad to put it on and… get it up and running."

Come the 14th of February, one day before the official unveiling to that high-ranking military audience, the demonstration worked – except the virtual missile didn't stop when it hit the ground. It just kept on going deeper and deeper. Fortunately, overnight Holberton – and what Bartik describes as her "nighttime logic" – worked out that a single switch was set wrongly and needed to be flipped. This corrected, the demo worked perfectly.

[28] Arthur Burks, 'Who Invented the General-Purpose Electronic Computer?', 2 April 1974, talk given at The University of Michigan, **rpimag.co/burkstalk**

[29] Interview with Jean Bartik by Gardner Hendrie on 1 July 2008, Computer History Museum, p28. Retrieved from **rpimag.co/bartikinterview**

Unlike the earlier press demonstration, there was an element of theatre this time too, as Mauchly and Eckert had painted numbers onto the bulbs. "[The] people that were sitting there could see the numbers build up as the shell reached altitude and then came down and hit the ground," explained Bartik. And it was a triumph.

It should have been a moment of triumph for the women programmers, too, but for Bartik it was bitter-sweet. While everyone else went out for dinner – Eckert, Mauchly, Burks, dignitaries from the University of Pennsylvania and BRL, plus the all the attending colonels and generals – neither Bartik nor Holberton were invited. "We were sort of horrified," said Bartik. "We knew how important it was and felt we had done so much but we didn't [get invited] – and of course in history afterward nobody ever mentions us."

Just to make the pill more sour, Goldstine claimed that the "main calculation and the interrelationship between the various problems were prepared solely by my wife and me"[30] in his book *The Computer from Pascal to von Neumann*, first published in 1972. Even when interviewed 36 years later, this still riled Bartik. "Well, he lied. I mean, he never even mentioned in his book the ENIAC women; all he ever mentioned was our names and who we married, and he called me Elizabeth. Well, my name was never Elizabeth."

To the watching public, however, the launch of the ENIAC was an unadulterated success. And one that would set the tone of media coverage for decades to come. In her 1995 article for the IEEE Technology and Society Magazine, Dr C Dianne Martin studied 43 newspaper articles. "In bold headlines seen around the world, metaphorical images such as electronic brain, magic brain, wonder brain, wizard, and man-made robot brain were used to describe the new calculating machine to an awestruck public," she wrote.[31] This, she argued, set the misleading anthropomorphic tone that still lingers for computers.

One such article appeared in *The Des Moines Register*.[32] Although it led with the curious headline, "ENIAC All Set to Make Mathematics Lost Art", it went on to describe the computer as "the supersonic quiz kid. It's the poor man's Einstein…

---

[30] Herman H Goldstine, *The Computer: from Pascal to von Neumann*, p230
[31] Dr C Dianne Martin, Department of Computer Science, The George Washington University, 'ENIAC: The Press Conference That Shook the World', in *IEEE Technology and Society Magazine*, December, 1995
www2.seas.gwu.edu/~mfeldman/csci110/summer08/eniac2.pdf
[32] Jack Wilson, 'ENIAC All Set to Make Mathematics Lost Art', in *The Des Moines Register*, 15 February 1946, p1

ENIAC is the army's new robot brain." Others managed to squeeze all the clichés into their headlines alone, including The Fresno Bee with "Army's Robot Einstein Solves Years' Problems In Hours".[33]

Aside from Einstein references, the tone of most articles was awe. Awe at the ability of this electronic machine, awe at the fact the United States created it secretly in such a short time, and awe at the potential good it could do. Nor did the ENIAC's creators aim to dampen expectations. In one article, Mauchly is quoted as saying that high-speed computing could lead to all sorts of future possibilities, including "better transportation, better clothing, better food processing, better television, radio and other communications, better housing, and better weather forecasting."[34]

The latter was one of his great loves, and the passion that drew him into the idea of electronic computers in the first place. So Mauchly must have been gratified to see that, alongside cosmic-ray studies, wind-tunnel design and many other scientific applications, the ENIAC was indeed used to predict the weather during its ten years of active duty.

By the time it was switched off – at 11.45pm on 2 October 1955 – the machine had also received several useful upgrades. For a start, it had become possible to store programs for "standard trajectory problems",[35] while each function table "became available for the storage of 600 two-decimal digit instructions." Both these changes resulted in major time savings, helping to keep the ENIAC relevant in the face of new electronic computers.

Such improvements happened after the Moore School had released the ENIAC to Aberdeen, and after Eckert and Mauchly had left; they had little choice, as a condition for staying on at the university was that they would lose all rights to the ENIAC's patents. Patents that would eventually lead to a lawsuit that would mean Eckert and Mauchly were, in the eyes of the law, no longer the inventors of the world's first electronic computer: that privilege, the judge decided, fell to Dr Atanasoff for the ABC.

---

[33] Associated Press, 'Army's Robot Einstein Solves Years' Problems In Hours', in *The Fresno Bee*, 15 February 1946, p5
[34] Thomas J O'Donnell, 'Mechanical Mathematician "Brain Child" of Hopkins Man', in *The Baltimore Sun*, 15 February 1946, p28
[35] Martin H Weik, 'The ENIAC Story', in *ORDNANCE*, Vol 45, No 244, January–February 1961, **jstor.org/stable/45363261**, p575

This may be surprising. It should be obvious that the ENIAC represents a huge leap over the ABC, to the extent that Dr Atanasoff didn't recognise any trace of his desk-sized computer on his first encounter with the machine in the late 1940s. Indeed, he claimed in a 1972 interview[36] that when he saw the ENIAC for the first time, "I looked at the ENIAC and said, I'm not interested in having any connection with it… Because it wasn't a very effective machine and I didn't like its end results and I didn't like the insufficiency and I didn't like this and that about it."

It was only in the late 1960s, he states in the same interview, that he read the ENIAC patent and "realised that Mauchly had taken ideas which I had devised and used them in the construction of the ENIAC patents." We cover this in more detail in Chapter 1, dedicated to the ABC.

However, the fact that a legal battle gave so much credit to Atanasoff should not diminish the work done by Mauchly, Eckert, and everyone else involved with the ENIAC. Atanasoff and Clifford Berry never took their ideas beyond a semi-working model that could solve a single type of problem; to create the ENIAC, a computer that continued to put in solid service for a decade, was a truly incredible achievement.

And let's follow in the ENIAC's footsteps and talk numbers. Mauchly and Eckert's machine used almost 18,000 vacuum tubes compared to 300 for the ABC. It used 20 accumulators rather than a pair of magnetic drums. And, although it is difficult to draw a direct comparison, the ABC could perform one complete operation every 15 seconds while the ENIAC could finish 5000 arithmetic instructions per second.

The key difference between the two machines, however, is that the ENIAC was a general-purpose computer. In other words, it could be programmed, unlike a special-purpose computer such as the ABC that could only solve one type of problem. And it had a much bigger legacy.

We're not simply referring to the computer that came afterwards, the EDVAC. By building the ENIAC, the Moore School essentially won the war over analogue devices such as the differential analyser. It proved that digital computing was the way forward, that it could solve problems that were previously unsolvable, and in one swoop moved us into a new, digital era.

---

[36] Interview with John V Atanasoff and Alice Atanasoff by Bonnie Kaplan on 17 July 1972, Smithsonian National Museum of American History, **si.edu/media/NMAH/NMAH-AC0196_atan720717.pdf**

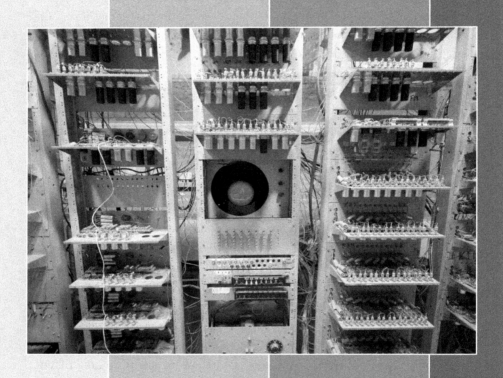

Manchester Baby replica
at the Science and Industry
Museum, Manchester, UK

Image: Parrot of Doom,
CC-BY-SA 3.0

Manchester Baby

"Computers were in the air"
FC Williams

There is something quintessentially British about the Manchester Baby. A plucky story of the underdog making something amazing happen despite the odds. A collection of donnish university professors who provided the great ideas while never seeking the limelight. And a certain amount of polite backstabbing when others were given too much public credit for creating the world's first stored-program computer.

But it's only right that we start the story with Professor Frederic Calland Williams, whose name was often shortened to FC or, to close friends, Freddie. As with so many other people featured in this book, he deserves to be called a genius. But unlike Alan Turing or John von Neumann, his genius lay in the world of engineering rather than mathematics. Although a prodigious academic researcher, with 20 papers to his name by the outbreak of war in 1939[1] (aged 28), he was at heart a problem-solver.

During World War II, those problems largely involved radar. Countless British pilots owed their lives to Williams's work on a system for distinguishing friend from foe in the skies,[2] and he was soon put in charge of automatic radar systems that helped fighter pilots work out the best route to intercept the enemy in the Battle of Britain and beyond. "He was most prolific, enthusiastic, and unselfish in his creativity," wrote James Whitehead, who worked with Williams during the war.[3] But he was also "notorious for his tangled breadboard circuits which often drooped over the edge of the bench towards the floor – a unique mix of conceptual elegance and material chaos."

This "material chaos" took place at the Telecommunications Research Establishment (TRE), discreetly tucked away in Malvern, Worcestershire.[4] But his reputation spread far wider. By the end of the war, knowledge of his expertise in radar and waveforms had spread to the Massachusetts Institute of Technology (MIT), which asked Williams to edit two volumes of its 24-volume opus on electrical engineering.[5] As part of his research in 1946, Williams headed over to

---

[1] T Kilburn and LS Piggott, 'Frederic Calland Williams', in *Biographical Memoirs of Fellows*, Royal Society, Volume 24, 1978, p584
[2] Williams was responsible for the Allied Forces' standard IFF Mark III (IFF stands for identification friend or foe), which marked a significant improvement on Mark II. Mark III was introduced to Allied aircraft, ships, and submarines from 1942.
[3] T Kilburn and LS Piggott, 'Frederic Calland Williams', in *Biographical Memoirs of Fellows*, Royal Society, Volume 24, 1978, p590
[4] This lies around 40 miles south-west of Birmingham, so was thought to be a safe place for vital radar research during the war.
[5] Mary Croarken, 'The beginnings of the Manchester Computer Phenomenon: People and Influences', in *IEEE Annals of the History of Computing*, Vol 15, No 3, 1993

America and discovered they were attempting to store information on cathode ray tubes (CRTs).

"It started by a visit of Freddie Williams to Bell Labs, where they were trying to get rid of ground echo on a radar by transferring the radar signal which occurs at the beginning of a trace on a ranging trace from nearby hills and buildings to allow any approaching plane to be seen," said Tom Kilburn, co-inventor of the Williams-Kilburn tube.[6] "Otherwise, the pulse from the incoming plane merges with the background, which is called the ground echo."

Bell Labs had the idea that if you could transfer this ground echo to a second CRT, it could be subtracted from the live radar image to leave only the radar signal for the plane. What lifted this beyond a simple improvement to radar was that the researchers were essentially devising a way to store information on a CRT.

Such work at Bell Labs wasn't happening in isolation. Historian Jack Copeland uncovered evidence that Williams "saw some work on nonregenerative CRT storage while he was at the Moore School" in June 1946.[7] And, while Williams didn't attend the Moore School lectures, it was also a feature of those given by Presper Eckert – co-creator of the ENIAC – that summer. Indeed, Eckert was so convinced that Williams based his ideas on what he saw at the Moore School that he disputed (unsuccessfully) Williams and Kilburn's American patent application.

Why the fuss? The big advantage of CRT storage over mercury delay lines (as used by the EDSAC and Pilot ACE, for instance) was that you didn't need to wait for data to appear. The information you needed was sitting there, ready to be accessed whenever it was needed. By comparison, mercury delay lines are sequential by nature, so data emerges in sequence from a queue. You had to wait for a bit of data to pass through the entire delay line before it could be read.

There are clear parallels with modern-day technology here. CRT storage was the first random access memory, now shortened to RAM, while mercury delay lines are akin to hard disks where data is stored on circular platters covered in magnetic

---

[6] Interview with Tom Kilburn by Geof Bowker and Richard Giordano, in *IEEE Annals of the History of Computing*, Vol 15, No 3, Jul–Sep 1993, p20

[7] Jack Copeland, 'The Manchester Computer: A Revised History. Part 1: The Memory', in *IEEE Annals of the History of Computing*, Jan-Mar 2011, p14. It's likely that Williams saw Robert McConnell's experiments with placing a metal plate over a CRT, a key part of the Williams-Kilburn tube design.

material; to minimise delays, these spin at ferocious speeds so that data can be read (and written) each time it passes over a recording head.

It's unclear if Williams started work on CRT storage immediately on his return from America or if someone asked him to research it. All we can now do is rely on Williams's own direct words: that "nobody was going to care a toss about radar" in post-war Britain, so he realised that he and others like him "were going to be in the soup unless we found something else to do. And computers were in the air. Knowing nothing about them, I latched onto the problem of storage and tackled that."[8]

First, he set up the same problem being tackled at Bell Labs: duplicating analogue signals from one tube to another. The two-tube setup never worked well enough to be reliable, but Williams was inspired to store data on a CRT tube as a gap in a line. So, a beam of electrons would streak across the phosphorescent material in the CRT to create a line, except for a tiny gap created when the beam would switch off before switching on again. This would show as a gap in the line, so a single bit of information.

This was a big step, but it wasn't the real breakthrough. What was: when Williams realised the act of switching off the beam left a tiny marker on the phosphor layer. A marker that told any subsequent beams where to switch off the signal. He called this the "anticipation pulse", successfully patenting the idea in November 1946.

In the intimate circle that formed British computing at the time, news of his success spread quickly. Keen to see Williams's work for himself, Sir Charles Darwin – grandson of the legendary naturalist and director of the NPL (National Physical Laboratory) – made the journey from London to Worcestershire for a hands-on demonstration.

This meeting was soon followed by another, this time at the NPL headquarters and attended by Turing, Williams, John Womersley, and Albert Uttley from the TRE.[9] The director of the TRE, Robin Smith, was also there, and at pains to explain that the TRE wasn't in a position to develop this specifically for the ACE project. The NPL then tried to lure Williams with a contract to "develop an electronic storage tube for A.C.E. machine" and "develop components of the arithmetical organ of the machine, e.g. adding circuit and multiplying circuit."

---

[8] Jack Copeland, 'The Manchester Computer: A Revised History. Part 1: The Memory', in *IEEE Annals of the History of Computing*, Jan-Mar 2011, p9

[9] These visits are detailed in Mary Croarken's article, 'The Beginnings of the Manchester Phenomenon: People and Influences', in *IEEE Annals of the History of Computing*, Vol 15, No 3, 1993, p12. Robin Smith's full name was Robert Allan Smith, but he was most often called Robin.

The CRT output of the Manchester Baby replica at the Museum of Science and Industry, Manchester
Image: Ben Green, Public Domain

It's easy to see why the NPL was so keen to bag Williams's services, but he had already accepted a position at the University of Manchester as Chair of Electrotechnics (Williams soon changed the department's name to Electrical Engineering). It was also a natural choice, as this was where he had graduated and earned his doctorate. Williams knew he would have far greater freedom, not least because TRE – a keen backer of the CRT storage concept as it had military applications – was willing to keep supplying all the materials he needed and sponsor an assistant to work with him.

The University of Manchester was equally determined to hire Williams. Max Newman, who played a key role in the creation of Colossus at Bletchley Park, had joined Manchester to head up its Mathematics department after the war. Earlier in 1946, he applied to the Royal Society for funding to build a computer laboratory at the university. Including, naturally, its own computer. This application almost fell at the first hurdle, because one of the five members of the deciding panel was Sir Charles Darwin. And he was not happy at the thought of a rival computer to the ACE.

This was despite Newman being at pains to draw a virtual line between his computers and the proposed NPL computer. "The object of the laboratory would be to produce pilot models, not to run machines on a production basis," his application stated. "Once a machine was running well, the time would have come to start a new one." He went on to explain that these would be used to solve academic problems such as the Riemann zeta hypothesis and algebraic theorems. His lab, in short, would "deal with mathematical problems."

Despite these statements, Darwin's objection was stiff. "It is not admissible to embark at this present time on two so similar machines, and that the prior claim should be given to NPL since they have the appropriate staff."[10] He also pointed to a lack of resources that the two computers would have to fight over, before firing a final salvo that the cost of Newman's lab "has been very seriously underestimated."

The Royal Society called in two referees to make a judgement: Douglas Hartree, at that point Plummer Professor of Mathematical Physics at Cambridge but previously a professor at Manchester University, and Patrick Blackett. The latter was always going to side with Newman, for not only was he Langworthy Professor of Physics at Manchester, but also best man at Newman's wedding (and, for that matter, Hartree's).

They both put forward a robust defence of Newman's plans, with Hartree emphasising the difference between a computer built for research, as proposed at Manchester, and the general-purpose computer being developed at the NPL. "Once one had found how to do one kind of problem, one would probably take the equipment to bits and try to do something else," he wrote.[11]

Despite the Royal Society calling in two further referees, Darwin found himself alone in his criticism of Newman's plans and withdrew his opposition in May. The next month, a smile must surely have spread across Newman's face when hearing that the UK Treasury (the Royal Society would administer the grant but did not hold the money itself) had awarded him £35,000 across five years. £20,000 to cover the computer's construction, £3000 per year to be spent on staff. In today's money, 2025, that equates to roughly £700,000 for the computer and over £100,000 per year in salaries.

---

[10] The original quotes come from Darwin's letter to the Royal Society on 20 February 1946, as quoted in Simon Lavington's article 'Early Days of Computing at Manchester: Max Newman's Royal Society Project, 1946-1951', 18 May 2022, in *IEEE Annals of the History of Computing*, Vol 44, Issue 2, p22

[11] Quotes taken from Hartree's letter to the Royal Society on 12 March 1946. This is quoted in Simon Lavington's same article as above, p22

So the University of Manchester's Royal Society Computing Machine Laboratory was born. By this time, Newman had hired mathematicians Jack Good (who worked with Newman at Bletchley Park) and David Rees, with the idea that they would spend half their time on the computer project. Their first job was to find out exactly what was happening elsewhere. This involved Newman and Good spending a week with Alan Turing to gain a better understanding of the ACE project, while Rees attended the Moore School lectures at the University of Pennsylvania. The same lectures that inspired Maurice Wilkes to build the EDSAC.

Newman was also in contact with John von Neumann and aware of his project to build a computer at the Institute of Advanced Study (IAS). "What I should most like is to come out and talk to you," he wrote in February 1943. Later adding: "I also hope to get hold of a good circuit man, though they are both rare and not procurable when found."[12]

This highlights Newman's problem: even if he had a clear vision for his computer, he was in no position to build one. He needed an engineering virtuoso in the mould of Colossus creator Tommy Flowers[13] to bring his hazily defined creation to life, but Flowers was tied up in post-war projects at the General Post Office. The GPO had initially committed to engineer and manufacture Newman's machine, but as with the ACE project this promise proved impossible to fulfil.

Newman's cause was dealt yet another blow when Professor Willis Jackson, who headed up the University of Manchester's Department of Electro-technics, left for Imperial College. Not only did he take his research group with him but also, the story goes, half the contents of the department's labs. We don't know exactly what equipment, but it may well have included components from Colossi as Newman had requested many of the bigger items – and ones that would not give away their origin – to be sent to Manchester after Churchill's order to dismantle them.

Which brings us back to why Blackett pushed for Williams to join the university, luring the gifted "circuit man" – to use Newman's phrase – with the vacant professorship. Even better that Williams would be bringing another gigantic brain with him in the form of Tom Kilburn, who we quoted earlier.

---

[12] Letter from Max Newman to John von Neumann, 8 February 1946, alanturing.net/turing_archive/archive/m/m14/M14.php

[13] Tommy Flowers was the engineer who came up with the idea of a rapid analysis machine, that would later be called Colossus, as we cover in Chapter 4.

Kilburn had graduated from the University of Cambridge with a first-class Maths degree in 1942, but on signing up for war duty was told to take a "take a City & Guilds crash course in electricity, magnetism, and electronics".[14] At which point he was despatched to Malvern and assigned to Williams's group. Their partnership took time to warm up. "I didn't know Freddie Williams until that day and in effect he said, 'Oh God, you don't know anything?' and I said, 'No'. That was the sort of relationship at the start."[15]

But the two men – both from northern England, with Williams born in Stockport (near Manchester) and Kilburn in Dewsbury, Yorkshire – would soon warm to each other. And produce excellent work, both at the TRE and the University of Manchester. Still, as Simon Lavington recalls, Kilburn would always be the junior member of the partnership, despite becoming a professor himself and founding the university's Department of Computer Science in 1964.

"I was on a very interesting train journey from Manchester down to London in 1976, for the opening of the Science Museum's first gallery of computing," said Lavington. "So there was Tom and FC and I on the train. It was very clear to me that FC was the senior person, and Tom was looking up to him." This conversation also showed that Williams's mind remained as inquisitive as ever. "We discussed various topics and one particular one was leaking window frames in Tom's house. And you could see FC's inventiveness getting to work about how to divert little runnels of water and stuff. His mind was always on the lookout for inventing things."

Williams would need all that inventiveness as he attempted to rebuild the Electronic Engineering department after the sudden departure of his predecessor.

"Now for three months or so, largely at the beginning of 1947, I did experiments varying the speed of the trace and the focus of the trace and so on," wrote Tom Kilburn in 1990.[16] "I came to the conclusion that another pulse, which didn't play any part in the original anticipation pulse, was much more useful than the anticipation pulse itself; and so sometime in March 1947 I convinced [Williams] that we could drop the anticipation pulse and use the positive pulse."

---

[14] Biography of Tom Kilburn on 'Digital 60 Manchester: 60 years of the modern computer', 2008, which is now archived at **rpimag.co/tomkilburnbio**

[15] As above

[16] Tom Kilburn, 'From Cathode Ray Tube to Ferranti Mark I', in *Resurrection: The Bulletin of the Computer Conservation Society*, ISSN 0958-7403, Volume 1 Number 2, Autumn 1990, **rpimag.co/fromcrt**

From here, Kilburn made quick progress, but he was not alone. Part of the deal with TRE was that they would supply a helper on secondment. Arthur Marsh wasn't enamoured with the work or the Manchester weather, so left after around three months and was replaced by Geoff Tootill in June 1947. Tootill, who had worked under Williams at the TRE, was lured by the prospect of earning an MSc on the job.

Lavington paints a picture of how the three men would work together, where Williams was very much in charge of the overall picture. "Of the three of them, the genius designer was undoubtedly Williams. No doubt Tom and Geoff, they were continually learning from him." And while Williams had too many other duties to spend whole days in the lab, he was always available if there was a problem to solve: Tootill said Williams would spend "an average of an hour a day, sometimes more and sometimes not at all, depending on the demands of his job as professor" in a 2010 interview.[17]

Together, the trio made incredible progress. By December 1947, Kilburn stated in his annual report to the TRE, they could store an incredible 2048 bits of information onto a single CRT.[18] This appeared in a 64×32 array and could be written to and read from in 0.2 seconds.

Before we move onto the creation of the Baby itself, it's worth pausing to reflect that whilst the Manchester team was making storming progress, the path of the contemporary American project – the Selectron – wasn't running so smoothly. The Selectron was being developed by the Radio Corporate of America (RCA) group in New Jersey. "It came out in a 256-binary-digit-per-tube version and was a work of great engineering virtuosity," wrote Herman Goldstine.[19] "Unfortunately, it was very complex in its structure and required technologies that were perhaps ahead of its time."

The Williams-Kilburn tube wasn't as sophisticated, but its "great practical advantage [over the Selectron] was that it could be made from off-the-shelf components and was consequently cheaper per stored bit," said Lavington.[20] Plus,

---

[17] Geoff Tootill interviewed by Thomas Lean, National Life Stories: An Oral History of British Science in partnership with British Library, C1379/02, track 3, recorded 8 January 2010, p08
[18] Tom Kilburn, 'A Storage System For Use With Binary Digital Computing Machines, 5.2 Storage Capacity of a Single C.R.T.', 1 December 1947
[19] Herman H Goldstine, *The Computer from Pascal to von Neumann* (Princeton University Press, 1993 paperback edition, ISBN 978-0691023670), p309
[20] In handwritten note to author

they were proven, commercial technology that were known to be reliable; the Selectrons were novel designs with all the unknown complications that brings.

Now that the Manchester team had a working storage tube, they needed to put it to the test. There was "no point just making a standalone test rig," added Lavington. "Better to build a complete small system using the Williams-Kilburn tubes and demonstrate real programmes running successfully for long periods." So the team started working on circuit designs for what was to become the Small-Scale Experimental Machine (SSEM), or Manchester Baby.

Of course, you can't just invent a computer out of thin air. Here, Williams gives credit to Newman. "Now let's be clear before we go any further that neither Tom Kilburn nor I knew the first thing about computers when we arrived in Manchester University," said Williams in a 1976 interview.[21] He adds that while Newman had funding from the Royal Society, as a mathematician he was not the "right sort of person to build a computer. So it was a very fruitful opportunity for collaboration between the maths department and the electrical engineering department – and Newman explained the whole business of how a computer works to us."

Kilburn's take, in a 1993 interview,[22] is less inclined to give credit to the mathematicians. In 1947, Kilburn recalls, he "attended some lectures which were given by Turing at NPL along with other people like Wilkes and so on... The only thing I got from this lecture was an absolute certainty that my computer wasn't going to look like that. Between early 1945 and early 1947, in that period, somehow or other I knew what a digital computer was, I knew how I would build it, and I knew how I wouldn't be building it."

He added: "There's not very much to learn... All you needed to know then is that the computer has a store which is alterable, that it goes through a program in order, and that it does its computing in an arithmetic unit. You don't need to know anything else. Right? Where I got this knowledge from, I've no idea."

Lavington believes Kilburn's most likely source of information was from the Princeton IAS project. Newman spent several weeks at Princeton learning about

---

[21] FC Williams interview with Chris Evans, 'The Pioneers of Computing: An Oral History of Computing', Science Museum (copyright Science Museum), 1976. The tapes were unavailable during research for this book, so this quote is taken from 'The Manchester Computer: A Revised History. Part 2: The Baby Computer' by Jack Copeland, 18 May 2022, in *IEEE Annals of the History of Computing*, Vol 44, Issue 2, p22

[22] Interview with Tom Kilburn by Geof Bowker and Richard Giordano, in *IEEE Annals of the History of Computing*, Vol 15, No 3, Jul–Sep 1993, p20

the proposed computer in late 1946, and it's likely that he shared that information with the university's new recruits. Uttley at the TRE also sent a description of the computer to FC at around this time.

However, there's a gulf between theory and practice. When it comes to allocating credit for designing the Manchester Baby, the intellectual heavy lifting – the overall design, the circuit diagrams, the inevitable problem solving – was done by the team of engineers.

This isn't to say that the Mathematics Department had no input whatsoever. In May 1947, for instance, Jack Good wrote an eight-page report entitled 'The Baby Machine' that included his suggestions for twelve instructions that could be included in a small-scale computer. Some have criticised this report as a mere simplification of previous suggestions given by von Neumann, but there's nothing wrong with being a conduit for the latest information flowing from the USA.

Ultimately, it was Kilburn who devised a prime factor program that would put the storage tube through its paces and a minimal IAS-style instruction set suitable for coding it up. Remarkably, he trimmed this down to a mere seven instructions in the SSEM. Even addition lost out: Kilburn realised that you include addition by using subtraction instead (turn the number you wish to add negative, then subtract it).

Construction of the Baby took place in a grimy lab previously dedicated to magnetism – it still had an enamel plate bearing that word in late 1948 – and, due to its central Manchester location, smuts of sticky black soot, spat out by nearby factories, would drift into the room. But, once the Baby grew in size, they couldn't close the windows because it produced so much heat that it would have been unbearable. This led to one memorable occasion where a squall sent rain through the open windows onto the hot valves, with predictable results.

While Williams, Kilburn, and Tootill produced the circuit diagrams, the building and wiring of the boards was mainly left to wiremen and technicians. This was a gradual process, as each time a board was wired it needed debugging. "We would screw it into a Post Office rack and connect it up to all the other units and then start to find out why it wasn't working properly," said Tootill. "Reasons for not working properly are classified into design errors and construction errors, and design errors were by far the most frequent."

Tootill remembers two wiremen – wirepeople, perhaps – who worked on the Baby. "[First] was Norman. I've forgotten his surname… he got promoted in the university's hierarchy and he was replaced by a young woman, Ida Fitzgerald, I remember her name, who was extremely efficient. She understood what was required from the circuit diagram and she executed the thing correctly and she did it very quickly too. [We] congratulated ourselves on being so lucky to have Ida working for us." So much so that they gave her the nickname Fabulous Ida.

The final credit in the creation of the Baby should go to the TRE, for it never stinted when Williams or Kilburn asked for more supplies.

Within a few months, the Baby took form. Its central components were three Williams-Kilburn tubes, with the first acting as its data store. This consisted of 32 rows of 32 bits (zeroes or ones), giving a grand total of 1024 bits – that is, one kilobit. The second CRT contained a 32-bit accumulator, where arithmetical operations took place. Then a third CRT held the address of the current instruction and the instruction itself.[23] None of these CRTs was visible to an observer, being covered by a metal plate as part of the tube's design. However, a fourth CRT was, and this could show what was currently being held on any of the three other CRTs. Effectively, that CRT was the Baby's output.

For input, programmers used a "keyboard of 32 buttons plus manual switches; these could be used to set any bit pattern in any word".[24] Buttons, incidentally, that were identical to those used in wartime planes such as the Spitfire, as the TRE drew upon the same stores as the RAF (Royal Air Force). Each instruction was limited to 32 bits, and in this tiny space the operator had to include the function code and the address of the operand (that is, the number that was to be operated on). It took around 1.2 milliseconds for each instruction to execute.

Let's not lose sight of the fact that the whole idea of building this mini computer was to ensure that the Williams-Kilburn tube could remain stable and operational in a 'real' environment, with refreshes happening at rapid-fire speeds. It was almost incidental that by performing such tests, the Manchester team were breaking such revolutionary new ground: this was the world's first stored-program electronic computer.

---

[23] These details are paraphrased from 'The Birth of the Baby' by Professor Hilary J Kahn, Dr RBE Napper, Department of Computer Science University of Manchester, IEEE Proceedings 2000 International Conference on Computer Design, September 2000, **ieeexplore.ieee.org/document/878326**
[24] As above, p482

Thankfully, the world's first 'stored program' is quietly impressive. Written by Kilburn, it found the highest factor of a given number ('a'), by simply dividing it by 'b'. In the first operation, b's value was a's value minus one; if that didn't produce a whole number, b's value became a's value minus two. And so on, until it found a number that divided exactly into 'a'. If we take the example of 'a' being 100, the Baby would cycle through 99, 98, 97, etc. until it reached 50 and, bingo, it had found the highest factor of 100.

While that calculation is trivial, when it came time to demo the Baby they were using 2^{18}. That tested 130,000 numbers over the course of 52 minutes, involving about 2.1 million instructions and roughly 3.5 million store addresses. An excellent test for the nascent storage tube technology.

The program ran successfully for the first time (albeit with a much smaller value for 'a') on 21 June 1948. But rather than call the newspapers, pop Champagne corks, or dance a merry jig, the team treated themselves to lunch in the canteen rather than their usual sandwiches.

The first the wider world knew of this development came in a letter to *Nature* magazine that was written in August but published on 25 September 1948. "A small electronic digital computing machine has been operating successfully for some weeks in the Royal Society Computing Machine Laboratory, which is at present housed in the Electrical Engineering Department of the University of Manchester," wrote Williams and Kilburn.

"The machine is purely experimental, and is on too small a scale to be of mathematical value ... However, apart from its small size, the machine is, in principle, 'universal' in the sense that it can be used to solve any problem that can be reduced to a programme of elementary instructions; the programme can be changed without any mechanical or electro-mechanical circuit changes."

By the time that letter appeared, the department had two new members: Dai Edwards and Tommy Thomas. Edwards, interviewed for the British Library in 2010,[25] painted a vivid picture of what met his eyes when he saw the Baby for the

---

[25] Professor David (Dai) Edwards interviewed by Thomas Lean, tape three, 5 March 2010, library shelfmark C1379/11/01/01-08 2010-02-26, 2010-03-05, 2010-03-12

Full view of the Manchester Baby, a composite of 20 photos taken by Alec Robinson on 15 December 1948; for an annotated version, visit **history.cs.manchester.ac.uk/?action=getSet&set=SET0002**
Image: courtesy of the University of Manchester; Alec Robinson married Sylvia Wagstaff, who was secretary to the Computer Machine Laboratory and who typed Turing's correspondence (which is held in the University archives)

first time. "What a mess," he laughed. "I mean it didn't look smooth and engineered, you know."

Visitors would have seen a collection of tall racks full of chassis, with most of the chassis packed with eight valves. And they needed a lot of power, Edwards explained. "So if you had a rack full of equipment, ten chassis, if they're all full of big valves that was eighty amps so how did you distribute all this power, just for heating the valves?" The answer is with great difficulty, which is why they kept the windows open in all conditions.

Edwards and Thomas were both brought on as research students, with Edwards contributing improvements to the storage CRT and Thomas focusing on slower magnetic storage. The latter was necessary for an improved version of the Baby, eventually called the Manchester Mark 1, to expand its storage capacity. They were soon joined by a PhD student, Alec Robinson, who created the Mark 1's multiplier, Colm Litting who worked on solid-state materials, and Cliff West, a servo-mechanics expert.

With the help of technicians who built the boards to the team's designs, the Baby was soon developing into a fully-fledged computer. There was also more memory, as it was relatively easy and inexpensive – compared to mercury delay lines or electromechanical relay switches – to add CRTs. But the focus, ironically given the importance of the focus/defocus method in the Williams-Kilburn tube,[26] had switched to computers. And now others were interested.

"Oh, we had one or two [visitors] a week I suppose," Tootill said. "I remember once I was in the lab on a Saturday morning – Tom never came in on Saturdays because he had such a long commute from Dewsbury – and one of the physics profs [Blackett] walked in and, he knew me, he said, 'Ah, Tootill, I'm jolly glad to see you… I'm very glad to see you here, perhaps you could give…' – this was another very eminent visitor he had – 'give a demonstration'. Well of course I'd got the thing partially disembowelled and I was engaged in updating something, but miraculously I managed to clip a few things together and make it work after a fashion and gave the demonstration."

The eminent visitor turned out to be Sir Ben Lockspeiser, Chief Scientist at the Ministry of Supply (MOS) and, despite the delay, he was impressed. So impressed that on 26 October 1948 he wrote a letter to Ferranti, the Manchester-based electronics manufacturer, authorising it to build a computer for the MOS.[27] A letter that, in the space of one sentence, commissioned what some argue is the world's first commercial computer: "You may take this letter as authority to proceed on the lines we discussed, namely, to construct an electronic calculating machine to the instructions of Professor FC Williams."

The official contract was signed in February 1949 for a very precise £113,783, two shillings, and seven pence. That's roughly £3.5 million in 2025. As a small matter of curiosity, that's almost exactly twice the amount the British government spent on Babbage's failed difference engine, according to the Science Museum[28] (once £17,500 in 1830 is adjusted for inflation).

Meanwhile, work carried on to improve the Manchester Mark 1, which the production version – the Ferranti Mark 1 – would be based upon. By this time Alan Turing had joined the University of Manchester as Deputy Director of the Computing Laboratory, but surprisingly his biggest contribution had nothing to do with the logical side of the machine, but instead its practical side: to specify a

---

[26] Initially the tubes used a dot and a dash to represent a zero and a one. However, in Edwards's words, the team "later decided to use a focus spot and a defocus spot, which were completely symmetrical and circular." This also avoided interference issues where a dot may appear to bleed into the dash, and vice versa.
[27] Simon Lavington, *Early Computing in Britain* (Springer, 2019, ISBN 978-3030151058), p15
[28] 'Charles Babbage's Difference Engines and the Science Museum', 18 July 2023, **rpimag.co/babbagesciencemuseum**

teleprinter for input and output along with the programming required to make it work. He wisely – none of the engineers spoke highly of Turing's engineering skills – left the physical implementation to the likes of Edwards and Tootill.

Perhaps surprisingly, Max Newman, silent for so much of this story, did play a role. He was an occasional visitor to the lab, usually when invited down by FC Williams, but on two occasions he, Williams, Kilburn, and Tootill discussed what other registers (storage tubes) could be included other than the arithmetic register and control register. "As a result of [the second] discussion, the B-tube... appeared," said Kilburn.[29] "Before the discussion there wasn't a B-tube; after the discussion there was a B-tube."

Essentially, the B-tube is the first implementation of what we would now call an index register, albeit a basic implementation. The core idea was to avoid the laborious practice of modifying instruction addresses manually, which not only saves time but also reduces memory requirements. With memory in such short supply, this proved an excellent improvement, even if Kilburn is reluctant to give much credit to Newman (or any of them) for this invention. "In the sense that index registers are common in computers... yes, it was important. In the sense of was it difficult to think of or build, no, it wasn't. I mean, compared with the difficulty of building the cathode-ray store, it was a trivial exercise."

By April 1949, the expanded Baby – now referred to as the Manchester Mark 1 – was ready to solve its first big problem, searching for Mersenne primes (which take the form of '$2^n - 1$', so where n is 2 the Mersenne prime is 3 and where n is 3 the Mersenne prime is 7). Originally set up by Newman, Turing developed the program with assistance from Tootill, and it earns its place in the hall of fame because it ran error-free from the evening of 16 June through to the early morning of 17 June. An exceptional example of reliability for a computer in 1949.

That same month the Manchester Mark 1 made its public debut, with the *Sunday Express* declaring "Mechanical 'brain' will learn to play chess" in a short article on 12 June.[30] While correctly crediting Professor FC Williams as "leader of the research

---

[29] Interview with Tom Kilburn by Geof Bowker and Richard Giordano, in *IEEE Annals of the History of Computing*, Vol 15, No 3, Jul–Sep 1993, p25

[30] Unnamed reporter, *The Sunday Express London*, 12 June 1949, p5

team" that built this computer, it must have been galling to Kilburn and Tootill that the two other named people in the article were Newman and Turing.

The article gives most space to Turing's quotes, which have much resonance in the current era of artificial intelligence. "In time, which may be years, of course, it should be possible to perfect and adapt the brain [the computer] so that it is capable of acting without human instructions, even of making decisions itself. I don't even draw the line about it writing a sonnet on a sunset."[31]

The Illustrated London News devoted a two-page spread, with a panoramic photo of the Mark 1, stretching across the top half. Headlined "A marvel of our time: the 'memory' machine which can solve the most complex mathematical problems", it provides a more factual account of the machine complete with keys telling readers what each rack of equipment did. Oddly, it got its Mark 1 computers mixed up, deciding to call the computer the "Automatic Sequence-controlled Calculating Machine," which is another name for the Harvard Mark 1.

Tootill left the team to join Ferranti in August 1949. He worked on the logic design for the Ferranti Mark 1, and his "informal report on the design of the Ferranti Mark 1 computing machine"[32] suggests that the specification was agreed by November 1949 whilst Tootill was still there. He would leave after four months over a dispute about pay, but this had no ill effect on the university's relationship with Ferranti – Kilburn says "it was a dream working with them" on the Mark 1.[33] It no doubt helped that Alec Robinson joined Ferranti in a similar liaison role in April 1950.[34]

It still took Ferranti over a year to turn the plans into a finished machine, which was delivered to the university on 11 February 1951. And this wasn't a switch-it-on-and-start-working moment. Lavington describes "three or four months of installation and commissioning work"[35] that followed. He adds: "The computer was at last in a good state when it was officially unveiled at an Inaugural Conference from 9 to 12

---

[31] Here are the first four lines of a sonnet about a sunset that Copilot, primarily based on ChatGPT-4o, produced in April 2025: "At day's last breath, the light does softly wane, A gilded orb sinks o'er a crimson sea. In whispered hues of fire, the skies explain, The art of twilight's tender mystery."
[32] Based on footnote 51 of Simon Lavington's article 'Early Days of Computing at Manchester: Max Newman's Royal Society Project, 1946-1951', 18 May 2022, in *IEEE Annals of the History of Computing*, Vol 44, Issue 2, p32
[33] Interview with Tom Kilburn by Geof Bowker and Richard Giordano, in *IEEE Annals of the History of Computing*, Vol 15, No 3, Jul–Sep 1993, p28
[34] TE Broadbent, *Electrical Engineering at Manchester University* (The Manchester School of Engineering, University of Manchester, 1998, ISBN 0-9531203-0-9), p183
[35] Simon Lavington, *Early Computing in Britain*, p15

July 1951, attended by 169 delegates including 13 from overseas." Lavington quotes Bernard Swann describing the "very large attendance of scientists, government officials, and interested businessmen. It was clear that a revolution had begun..."[36]

It had. The Ferranti Mark 1 was the world's first commercially available computer,[37] being delivered to its customer – the university, which would manage the computer for the Ministry of Supply – several weeks earlier than the UNIVAC. But supporters of the latter's claim will point out that its opening ceremony happened a month earlier than the Mark 1's, on 14 June. But let's not bicker: what's most important to the world is that, on either side of the Atlantic, computing had expanded into new frontiers.

There was also a fun side to the new computer. Within a few years of the Mark 1's arrival, the walls of the university lab featured love letters written by the computer. "Darling Sweetheart," one began, "You are my avid fellow tiding. My affection curiously clings to your passionate wish. My liking yearns for your heart. You are my wistful sympathy: my tender liking."[38] Not a poem to bring a tear to the eye, perhaps, but its meaning is clear.

British computer scientist Christopher Strachey – a friend and colleague of Turing at the university – explained how he came to write the program that wrote this poem in a still-relevant article entitled 'The "Thinking" Machine' in 1954. He also explains how, two years earlier, he created what Guinness World Records considers the first video game: not chess but draughts (or checkers).[39] Addressing those who might be worried that computers were going to take over the world, he wrote: "It is impossible to over-emphasise the importance of the program; without it a computer is like a typewriter without a typist or a piano without a pianist."

Sharp-eyed readers may at this point wonder what happened to the £20,000 earmarked for the computer by the Royal Society. After all, the components for the Baby and the Manchester Mark 1 it became were all supplied by the TRE. And the Ministry of Supply paid for Manchester University's Ferranti Mark 1. The answer is that it went towards the building that accommodated the computer.

---

[36] Simon Lavington, *Early Computing in Britain*, p16
[37] See 'Milestones: Manchester University "Baby" Computer and its Derivatives, 1948-1951' for the details behind this claim, which has been accredited by the IEEE. **rpimag.co/manchesterbaby**
[38] Christopher Strachey, 'The "Thinking" Machine', in *Encounter*, October 1954, p26
[39] 'First videogame', Guinness World Records, **guinnessworldrecords.com/world-records/first-computer-game**

Throughout the story of the Baby, there is a feeling of Newman's original vision – and it was never truly defined – slipping away from him. Of events in the Electrical Engineering department overtaking his idea of mathematics-focused computers that he put forward in his 1946 paper to the Royal Society. But two years proved a long time in the early days of electronic computers, with once-promising technologies such as the Selectron falling by the wayside.[40]

However, it's clear that Newman was consulted along the way, and he had input into the Ferranti Mark 1. His work on the Mersenne primes, carried on by Turing, also indicates his interest in the machine. Plus, when the Computing Machine Laboratory (the Royal Society name had been quietly dropped) hired staff to provide programming support, it tended to be mathematicians, such as Cicely Popplewell and Audrey Bates who joined in 1949.[41]

Newman's interests drifted away from computers in the early 1950s, while Turing remained an enthusiastic user of them until his tragic early death in 1954. He even wrote the first programming manual for the Manchester Mark 1, but his habit for creating his own notation (and assuming too much knowledge on the part of the reader) meant it met with a dim response.

The Mark 1 was the only computer that FC Williams helped to design. He famously never used one either. "I'm not really interested in computers," he said in a 1977 interview for the British Science Museum.[42] "I made one and I thought one out of one was a good score so I didn't make any more."

Kilburn had a similar approach to programming – the demo program he wrote for the Baby remained his only contribution to the art – but for him the Mark 1 was just the start. He and the university would play a major role in designing four more commercial Ferranti Computers, including two that were based on designs created at the university as part of its research. In particular, the Atlas was based upon the MUSE project (1958–1962).

But perhaps the most meaningful of Manchester's computers was the Ferranti Mark 1* (aka Mark 1 Star). While based upon the Mark 1, it was a sharpened

---

[40] Eventually, the Princeton IAS computer – which became fully operational in June 1952 – would use the Williams-Kilburn tube, not the Selectron.
[41] TE Broadbent, *Electrical Engineering at Manchester University*, p183
[42] FC Williams interview with Chris Evans for the Science Museum, 1977. Independently transcribed by Jack Copeland and Simon Lavington.

version made at the behest of John Bennett. "He was in the Australian RAF service in radar during the war, he got interested in computing and went to Cambridge," said Lavington. That was in 1947, where he became the EDSAC team's first PhD student. On gaining his PhD, he joined Ferranti where he led a team of about 15 programmers.

"He also took the lead in specifying the Ferranti Mark 1*, which was much more successful than the Mark 1," added Lavington, whose book *Early Computing in Britain* covers the glory days of Ferranti's computing division. In particular, it details Bennett's proposals for a new computer, which he referred to as the Mark II in his document dated 21 March 1951, and which jettisoned mathematical-leaning instructions such as Turing's random number generator in favour of ones geared towards "engineering and physical problems, e.g. floating-point operations, etc.".[43] Crucially, it would be easier to program as the operation code was slimmed down to five bits (from six), making it directly compatible with the five-bit teleprinter codes then in universal use.

The Mark 1* would make its own mark on computing history when it was bought by Shell, as it appears to have been the first computer bought for purely commercial reasons (and with no government aid). Installed in Amsterdam during 1954, the MIRACLE (as Shell christened it) put in seven years of service, including one memorable night where it calculated the best route for its crude oil tankers when the Suez crisis of 1956 closed the canal.

Kilburn would continue to be pivotal in the wider success of computing at the University of Manchester. He founded the Computer Science department in 1964 – the first in the UK – and from its starting cohort of 28 students in 1965 it grew to 150 by the time he retired in 1981. During that time, the department continued to design and build its own computers.

While Ferranti's fortunes eventually foundered, along with the rest of the British computer-building industry, this conveyor line of computer scientists – not to mention early programming contributions such as the Autocode, which predates **FORTRAN** and is considered by some to be the first high-level language – is ultimately the Baby's legacy. But it was a University of Birmingham graduate, Chris Burton, who led the team to rebuild the Baby in time for its 50th anniversary in 1998. The replica now

---

[43] Simon Lavington, *Early Computing in Britain*, p79

lives in the Science + Industry Museum in Manchester, with hands-on demonstrations still happening thanks to volunteers.[44]

It took four years to build the replica, in part because they were determined to use genuine parts throughout, but just like the original it worked when it was time to be unveiled. Tom Kilburn's only criticism on seeing the final model: that it looked far too clean compared to the real thing.

---

[44] We can't guarantee you'll see a live demonstration if you just turn up! In April 2025, the website stated: "Please note: We normally try to demonstrate Baby on Mondays, Tuesdays, Thursdays, and Saturdays between 10.30–13.30, but we can't always demonstrate Baby at these times because of volunteer availability. For more information, speak to a member of staff or check the 'Today's events' sheet when you arrive."

Bill Renwick and Maurice Wilkes with the EDSAC

Image: copyright Department of Computer Science and Technology, University of Cambridge. Reproduced by permission

EDSAC

(Electronic Delay Storage Automatic Calculator)

In the running to create the first stored-program computer

Click-click-click. This was the inauspicious, mundane sound that marked the true beginnings of stored-program computing, as a thin thread of tape wound its way through a reader on 6 May 1949. Two minutes and 35 seconds later, the teleprinter spluttered into action, producing a table of squares from zero to 99.

And it all happened in Cambridge, England, 3500 miles and an ocean away from Philadelphia where the EDVAC was still being assembled.

By all rights, the EDVAC should have been the first stored-program computer. Presper Eckert and John Mauchly, the ENIAC's main creators, fully understood their first computer's limitations and started planning the EDVAC, its natural successor, in mid-1944. That was more than a year before the ENIAC became operational.

With mathematician John von Neumann helping to create a logical outline for this next-generation computer, the US Army giving its financial backing, and a team that was uniquely armed with experience gained from creating the world's first general-purpose electronic computer, the Moore School was surely in the perfect position.

But no. That honour would ultimately fall to Britain, with the Manchester Baby (see Chapter 7) acting as proof-of-concept and EDSAC following a matter of months later. There are numerous reasons for this, and many people to earn credit, but the giant among them is the pragmatic genius of Maurice Wilkes.

While Eckert and Mauchly were bogged down by patent disputes and their desire to set up a commercial computer company – leading them to leave the army-backed EDVAC project – Wilkes was a one-man decision-making machine. He also had the riches of the University of Cambridge to draw upon. And he was eminently practical.

Unlike his Cambridge contemporary Alan Turing, who would lead the ACE project (see Chapter 10), Wilkes's interests had always leaned to applied rather than pure mathematics. After graduating with first-class honours in 1934, he gained his PhD in 1937 through work at the University of Cambridge's famous Cavendish Laboratory, where he studied radio wave propagation.

In his tribute to Maurice Wilkes, published in the Biographical Memoirs of Fellows of the Royal Society,[1] Martin Campbell-Kelly writes: "The work involved excursions in a motor car, towing a caravan loaded with portable electrical measuring equipment, taking readings in the field, and undertaking a mathematical analysis

---

[1] Martin Campbell-Kelly, 'Sir Maurice Vincent Wilkes: 26 June 1913 – 29 November 2010', in *Biographical Memoirs of Fellows of the Royal Society*, Vol 60 (2014), **jstor.org/stable/24868297**, pp433-454

afterwards." Campbell-Kelly adds that in "every way Wilkes was in his element".

The mathematical work involved solving differential equations, so when Wilkes discovered that John Lennard-Jones – a professor of theoretical chemistry at Cambridge – had commissioned a model of Vannevar Bush's differential analyser, he asked permission to use it. Despite using Meccano parts for many of the components, this was no toy but a working tool; Wilkes rapidly became its master, publishing several papers using its results. The Meccano-based analyser played another role, too.

Maurice Wilkes and EDSAC
Image: copyright Department of Computer Science and Technology, University of Cambridge. Reproduced by permission

"This was a model intended for propaganda purposes so that [Lennard-Jones] could get the money to have a full-scale differential analyser built according to Bush's plans," Wilkes told an audience in 1979,[2] "but instead of just saying he wanted a differential analyser and a room to put it in, he said he wanted a computing laboratory." The university agreed, although they insisted on calling it the Mathematical Laboratory.

Lennard-Jones was appointed part-time director of the Laboratory in October 1937 with Wilkes its sole member of full-time staff, only for World War II to intervene shortly before the full-size differential analyser was installed. For the next six years, the Laboratory would be seconded to the war effort[3] while Wilkes devoted his time to radar research and installations.

For example, he worked out that by measuring the amplitude of radar echoes, you could measure the size of the vessel being tracked. He passed this information to operators of coastal radar systems who were then able to work out which types of vessel were being tracked. Before this, the equipment had even confused seagulls for boats.

Wilkes returned to Cambridge in the summer of 1945 to discover that Lennard-Jones wished him to take on the running of the nascent Mathematical Laboratory.

---

[2] 'The Birth and Growth of the Digital Computer, lecture by Professor Maurice Wilkes, 1979, Computer History Museum, **youtu.be/MZGZfsr1KfY**

[3] According to a PhD dissertation submitted by Mary Goretti Croarken in 1985, it was used by the Ministry of Supply. 'The Centralization of Scientific Computation in Britain 1925-1955', p83

Wilkes readily agreed, and soon outlined what he saw as the top priorities for the lab: to teach science undergraduates practical computational techniques and to provide a computing service. At this point, such a service primarily meant access to desk calculators and the differential analyser.

In one of the many strokes of good fortune that littered the EDSAC's creation, Wilkes had struck up a friendship with Douglas Hartree before the war. Wilkes had visited him in Manchester to see the university's full differential analyser in action. "He would go to any amount of trouble to help people," wrote Wilkes in his memoir.[4] "He also took me along to see the people at Metropolitan-Vickers, who had been responsible for the building of his machine and who were expecting to be responsible for the Cambridge one."

Unlike Wilkes, Hartree was central to British computing efforts during the war, and had seen the ENIAC in America long before its existence was made public. It was Hartree who fuelled Wilkes's interest in digital computing on his return to Cambridge, with talk not only of the ENIAC but also the work going on at Harvard.

Then, in May 1946, Wilkes was visited by Leslie Comrie, who had – with remarkable foresight – created a commercial computing company in 1936. This was naturally based on mechanical methods, but Comrie immediately saw the importance of the ENIAC. And most importantly, he lent Wilkes his copy of the 'Draft report on the EDVAC' for one night. Wilkes stayed in the office late into the evening so he could read it before Comrie left the following morning.

Time for another piece of good fortune. "An event which shaped the whole of my subsequent career occurred in the summer of 1946," Wilkes later recalled,[5] "when I received a telegram from Dean Pender of the Moore School of Electrical Engineering in Philadelphia, inviting me to a course on digital computers." Wilkes immediately accepted, but could not travel to America quite as quickly as he wished due to the British government retaining control of all shipping.

"Those days, it was common for cargo ships to have accommodation for twelve passengers," said Wilkes at EDSAC '99, an event that took place at the University of Cambridge to mark the 50th anniversary of the first program.[6] "I travelled on such a

---

[4] Maurice Wilkes, *Memoirs of a Computer Pioneer* (MIT Press, 1985, ISBN 978-0262231220), p29
[5] 'The Birth and Growth of the Digital Computer, lecture by Professor Maurice Wilkes, 1979, Computer History Museum, **youtu.be/MZGZfsr1KfY**
[6] Maurice Wilkes, 6 May 1999, EDSAC '99 conference, EDSAC disc 1, **rptl.io/edsacdisc1**

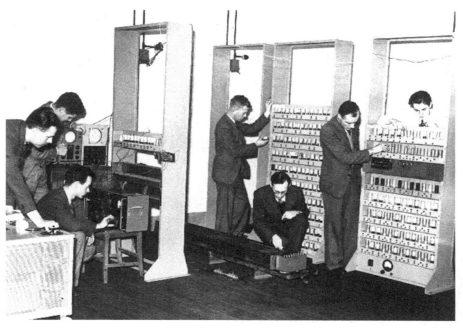

EDSAC under construction in 1947, PJ Farmer, R Piggott, MV Wilkes, W Renwick, SA Barton, GJ Stevens, JM Bennett
Image: copyright Department of Computer Science and Technology, University of Cambridge. Reproduced by permission

ship, but there were 36 passengers so we were a little crowded." These were all men, and the shipping company wanted to make sure no women were among them. "It is the only time in my life in which I had a medical examination directly to determine my sex."

The delays with transportation, plus problems with the ship's engines, meant Wilkes only arrived for the final two weeks of the eight-week course. Fortunately, the first six weeks covered areas he already knew well due to his extensive experience with differential analysers and knowledge of the EDVAC report. He devoured the lectures by John Mauchly and Presper Eckert, who were extremely generous in sharing their knowledge with the attendees. Mauchly even gave Wilkes a personal tour of the ENIAC and went through the construction details with him.

Through Hartree's contacts and invitations, Wilkes also had the chance to speak to other luminaries of the computing world. This included Howard Aiken at Harvard, who demonstrated the Harvard Mark I in operation and gave him a glimpse

Wilkes with mercury delay lines
Image: copyright Department of Computer Science and Technology, University of Cambridge. Reproduced by permission

of the Mark II, then under construction. He also visited Samuel Caldwell, who had helped Vannevar Bush develop the original differential analyser at MIT, and picked the brains of Herman Goldstine about the EDVAC's design over a dinner in Philadelphia.

"I felt at the end [of the trip] that I knew all that was to be known about this new field. And I gradually found myself gripped by the idea that, come what may, we were going to build a computer in Cambridge," Wilkes later wrote. In his final week in Philadelphia, and during the course of the return journey on the Queen Mary, he sketched out his early plans for what would become the EDSAC.

Wilkes knew that creating a means for storing data, the memory, would be the biggest challenge. Based on Eckert's suggestion, he didn't hesitate to choose mercury delay lines. These had become a vital component in World War II, where they were used to store sound waves from radar. In the EDSAC, each 'blip' fed into the line represented a one in binary, so each delay-line could store a number – a series of zeroes (no sound) and ones – that could be read at the other end. "They were cleaned up, amplified, resynchronised, and put back in at the beginning," said Wilkes at the EDSAC '99 event. "So these pulses went round and round and round and in that way the pattern of the pulses was preserved and stored."[7]

At this point, Wilkes had another stroke of luck, meeting Tommy Gold, who provided a design for a mercury delay line that he had helped to develop during the war.[8] Wilkes first built a single tank to Gold's specifications, although he skimped on the diameter so that less mercury was required. After all, he had to build 32 tanks for the final product.

---

[7] Wilkes provided a rather more technical description in a 1948 article called 'A Discussion on Computing Machines', published by the Royal Society of London. "The delay unit itself consists of a tube filled with mercury. The ends are closed by means of two similar X-cut quartz crystals. Electric pulses applied to one of the crystals gives rise to ultrasonic pulses which travel through the mercury with the velocity of sound. When they reach the far end of the tube, they are reconverted by the second crystal into electric pulses."

[8] Maurice Wilkes, 'The thinking behind EDSAC', in *Resurrection*, Autumn 1990, p7

9-inch cathode ray tubes (CRTs) used as monitors to display the contents of the computer's memory and other registers
Image: copyright Department of Computer Science and Technology, University of Cambridge. Reproduced by permission

Each tank, or delay line, was about 1.5 metres long, and could store 576 binary digits (bits).[9] The 576 bits in each tank was divided into 16 36-bit words. Each 36-bit word had one bit used to separate the words, and one bit that indicated the sign of the number (negative or positive). Across 32 tanks, that meant EDSAC could store 512 numbers, each with ten decimal digits.

The great thing for Wilkes is that he was in charge. While the EDVAC team had two masters to answer to, in the Moore School and the US Army, Wilkes had a clear remit, endless internal resources to draw upon, and no chain of command. "I didn't have to ask anybody 'could I build a computer, please?'. I didn't have to put it in any proposal. I didn't have to arrange any budget. I was in charge and I could go ahead," he said.[10]

Wilkes went into more detail in an article for the Computer Conservation Society in 1990.[11] "It was not a project to build a computer only. It was a project to build a computer, to learn how to use it, and then to solve some problems." And he

---

[9] An Ultrasonic Memory Unit for the EDSAC, Maurice Wilkes, Computer Conservation Society, **rpimag.co/edsacmemory**
[10] 'The Birth and Growth of the Digital Computer', lecture by Professor Maurice Wilkes, 1979, Computer History Museum, **youtu.be/MZGZfsr1KfY**
[11] Maurice Wilkes, 'The thinking behind EDSAC', in *Resurrection*, Autumn 1990, p4

Chapter 8: EDSAC   157

understood what sort of problems needed to be solved thanks to his regular dealings with students.

Nor was he concerned about creating the perfect computer. "[We] just barged ahead on the EDSAC and the rule was that if you had got something that would work, you didn't spend another hour on making it simpler or cheaper, you went ahead with it."

He also made one key pragmatic decision early on: to aim for a pulse rate of 500Hz (500 pulses per second) for the vacuum tubes rather than the 1000Hz that the EDVAC and BINAC used. Wilkes knew that this would mean his computer would be slower than rivals, but as the EDSAC would be so much quicker than what it replaced, he felt it was the best decision. It is certainly one of the reasons for the EDSAC appearing first.

Wilkes's practical side also came to the fore. Right from the start, he knew he would have to build parts himself and so the Mathematical Laboratory included its own instrument workshop. The EDSAC could not have been completed without the skills of instrument maker Gordon Stevens, for instance.

"We had to make things like tape readers," said Stevens at an event in 1999 marking the EDSAC's 50th anniversary,[12] "and modify various telegraph equipment so that you could get information into the machine. In EDSAC, all of the stuff that went into the machine was on telegraph tape."

While Wilkes was keen to dismiss the notion that EDSAC was built from "war surplus junk", he took every opportunity he could to use army surplus equipment, including the crucial valves. Much had to be built to specification, too, such as the two giant 'batteries' that would store the mercury delay lines and keep them at the required temperature. That job fell to the university's central workshop, part of its Engineering Department, while Wilkes outsourced construction of the chassis to a local firm.

By January 1947, the trio of Wilkes, Gold, and the lab's principal assistant, Philip Farmer, had managed to get the first mercury delay line working. From this point on, the team slowly grew and with it, despite continuing hardware shortages, momentum. Electronics engineer Bill Renwick arrived in March 1947, Richard Kimpton joined

---

[12] 'Behind the green door', University of Cambridge Collections, EDSAC 99, rptl.io/edsacgreendoor

EDSAC with an experimental magnetic tape rack (added in 1952) and a teleprinter for the output
Image: copyright Department of Computer Science and Technology, University of Cambridge. Reproduced by permission

straight from school that summer, before Vic Clayden and Herbert Norris rounded out the team.

A further boost came in July 1947 when Raymond Thompson and Oliver Standingford from the famous Lyons coffee house visited. This ultimately led to Lyons offering Wilkes a £3000 cash injection and the services of an assistant, with the idea that this extra pair of hands would learn on the job and aid Lyons with the building of their own copy of the EDSAC. Wilkes readily accepted.

Although Wilkes is rightly lauded as the creator of the EDSAC, he also drew upon the skills and knowledge of two University of Cambridge students. First, he delegated some key design work to research student John Bennett. "I was responsible for designing, constructing and testing the main control unit," Bennett wrote in 1999.[13]

"This unit sequenced the machine through the cycle of extracting from the store and decoding instructions (orders, we called them), extracting operands, initiating

---

[13] EDSAC 1 and after – a compilation of personal reminiscences, Dr David Hartley, University of Cambridge Computer Laboratory, April 1999, **rpimag.co/edsac1andafter**

individual arithmetical and logical processes, and proceeding to the next instruction. I also designed, constructed, and tested the bootstrap facility."

Bennett was joined by fellow research student David Wheeler in September 1948. And he made an immediate impact. Wheeler was responsible for the programming system, and proved instrumental not only in the success of EDSAC but also in the creation of programming concepts that would become hugely influential. He was the one who created the 'initial orders' for EDSAC, which loaded the instructions from the tape. He also formalised the idea of a closed subroutine, although it should be emphasised that John Mauchly deserves credit for the idea of subroutines (even if he did not call them as such).

Wilkes also relied on Bill Renwick who had joined in 1947 as a research assistant but became the de facto chief engineer according to Gordon Stevens, the EDSAC's principal instrument maker. He would die, far too young, in 1971, and was evidently much missed at the 50th anniversary in 1999. "He was a fine engineer and a fine man," said Stevens.[14]

Indeed, much of the credit for the EDSAC actually working should go to Bill Renwick. "I was what would nowadays be called the chief architect," said Wilkes.[15] "I also designed many of the early chassis and put them through their preliminary debugging. I then handed them over to Renwick who, as chief engineer, had the task of putting all these chassis together and making them work." Over time, Renwick would take over more of the design side too.

By the autumn of 1948, all the components were created: now, like Frankenstein's monster, everything had to be combined into one. Again, this job fell to Renwick. In February, they could read instructions from the input tape into the memory. Next step: attaching a teleprinter so that output could be read. Next, came the slow and agonising process of problem-solving, a portent of all the debugging of computers to come.

Finally, on 6 May 1949, came that magical click-click-click that began this chapter, followed by the chirruping of the teleprinter with the table of results. The EDSAC worked. The team had created the world's first stored-program, general-

---

[14] 'Behind the green door', University of Cambridge Collections, EDSAC 99, **rptl.io/edsacgreendoor**
[15] Maurice Wilkes, 6 May 1999, EDSAC '99 conference, EDSAC disc 1, **rptl.io/edsacdisc1**

purpose electronic computer. It says much about David Wheeler that his first response was to write a program to find prime numbers.

While there are good reasons to celebrate the 6th of May, it was by no means the end of the EDSAC's development. Its whole reason for being was to serve as a useful tool for the university, and Wilkes was keen for it to be as user-friendly as possible. For example, creating a subroutine to convert decimal numbers into binary so that students would not need to reinvent the wheel each time.

Sid Barton with the EDSAC 2 and its miniature valves
Image: copyright Department of Computer Science and Technology, University of Cambridge. Reproduced by permission

After a few weeks, Wilkes felt the EDSAC was ready for its first public demonstration: at a conference held in Cambridge on the apt topic of high-speed automatic calculating machines. Along with many from academia, attendees included representatives from Ferranti, Elliot and Lyons, a handful of UK army officers, and several interested parties from abroad – including France, Germany, Sweden, and the Netherlands.

The attendees were suitably impressed, but you will struggle to find much mention of it in the papers of the time. That's despite Wilkes and Renwick giving a personal demonstration of the EDSAC to a reporter from *The Daily Telegraph*, one of the UK's most prominent newspapers, which only found room for the computer's arrival on page 15 of its edition on Friday 17 June 1949.

"NEW 'BRAIN' STORES ORDERS" was the story's hardly compelling headline. "It has a 3500-valve 'brain' weighing about a ton," the unnamed reporter wrote, before alluding to some concerns (even more relevant today) that these electronic brains could take over the world. "Statements that have given almost human qualities to the machine, and alarmed some people, are misleading. Its marvel is in being able to read, subtract, multiply, and divide automatically at speed, completing in weeks what would take humans years."

Wilkes directly addressed this in his interview with *The Daily Telegraph*. "I liken it to a calculating machine operated by a moron who cannot think but can be trusted to do what he is told," he said. "If you tell the machine to do something silly, it goes on doing it."

He went on to explain the types of calculations he anticipated the EDSAC would solve, including nuclear physics, aerodynamics, and astrophysics – such as working out why stars are a certain size and brightness.

Aside from the PR exercise, Wilkes had more immediate, practical concerns. "By June 1949, people had begun to realise that it was not so easy to get a program right as had at one time appeared," he wrote in his 1985 book, *Memoirs of a Computer Pioneer*.[16] "It was on one of my journeys between the EDSAC room and the punching equipment [one floor below] that, hesitating at the angle of the stairs, the realisation came over me with full force that a good part of the remainder of my life was going to be spent in finding errors in my own programs."

Always keen to share their knowledge, Wilkes, Wheeler, and Stan Gill would compile their notes on programming the EDSAC – including a chapter on debugging – that was shared informally as a typewritten manuscript but would become the world's first programming book when published in 1951. Officially called 'The Preparation of Programs for an Electronic Digital Computer,' it was affectionately shortened to WWG to reflect the three authors' surnames.

By this time, the EDSAC was operating 24 hours a day, seven days a week, but like all the early computers it was far from reliable. A team of engineers was on hand to fix it during the day but "when they left in the evening, we programmers would continue running our programs until the next breakdown happened, when we would switch off and go home," reported D H Shinn, then a research student.[17]

That normally worked well, giving them a chance to get some rest. "However, there was an occasion, probably in March 1950, when EDSAC refused to break down," said Shinn. "It continued working throughout the night. When the engineers arrived in the morning... I was the only programmer left; the others had gone home to have their breakfasts; I soon went home to have my breakfast, and a good sleep."

The Cambridge team continued to improve the reliability and feature set of the EDSAC over its life, it remained a temperamental beast, delivering electric shocks to

---

[16] Maurice Wilkes, *Memoirs of a Computer Pioneer*, p145
[17] 'EDSAC 1 and after – a compilation of personal reminiscences', Dr David Hartley, University of Cambridge Computer Laboratory, April 1999, **rpimag.co/edsac1andafter**

one user in the mid-1950s. "The EDSAC 'console' was an old wooden table; when using the machine, one could sit with one's knees under it," remembered Jenifer Haselgrove (née Wheildon Brown, later Leech), a research student between 1953 and 1956.[18]

"One hot summer night I was working late, wearing shorts. Sleepily I reached out to the left to put a data tape in the tape reader, and was woken sharply by an electric shock. A little investigation revealed an unprotected rheostat, with mains voltage on it, under the table. My bare knee had been pressing against it."

So EDSAC wasn't perfect. Its 500-word memory also limited the problems it could tackle. However, as David Wheeler eloquently put it in a 1987 interview, "a machine selects the problems that they can do".[19] In EDSAC's case, he said, it selected "differential equations, Fourier transforms, and other things which were not quite so space-intensive."

This also ties in with a different computing philosophy at Cambridge compared to its American counterparts. Wilkes always intended EDSAC to be a tool for use by students at the university, and didn't want it to be tied up for weeks at a time on a single problem. He also wanted to create a generation of computer-literate graduates, and this legacy can be seen echoing through the years.

Cambridge, unlike its great rival the University of Oxford, provided defining moments of computing in the following decades. Many electronics and early computer companies – including Sinclair Research and Acorn, creators of the ZX Spectrum and BBC Micro respectively – based themselves in Cambridge, at least in part to benefit from the students who took Computer Science courses.

Some individual graduates are particularly noteworthy. Steve Furber and Sophie Wilson co-created the ARM architecture that almost every smartphone processor in the world relies on. Then there's Eben Upton, the CEO and co-creator of the Raspberry Pi, the best-selling computer in history[20]. Would this innovation have happened without Wilkes's foresight and drive? It somehow seems unlikely.

---

[18] As above
[19] An interview with David J Wheeler by William Aspray, Charles Babbage Institute, 14 May 2007, **rpimag.co/wheelerinterview**
[20] Les Pounder, 'Raspberry Pi celebrates 12 years as sales break 61 million units', Tom's Hardware, 29 February 2024, **rpimag.co/61million**

The EDSAC also had a direct descendent: what came to be known as EDSAC 2, with the original often referred to as EDSAC 1. Wilkes began thinking about it in earnest once the first machine's teething problems had been largely solved, and he was keenly aware that its successor needed to be both faster and more reliable.

Wilkes secured £25,000 of funding from the Nuffield Foundation in June 1951, by which time he had some clear ideas in his mind. One of which was a switch from serial operations to a parallel system. At that point, he kept his options open in terms of memory. Mercury delay lines remained the proven technology, but it was slow compared to the rest of the computer, so would be a bottleneck.

He hoped that a superior alternative would come to light during the development process, and this proved absolutely correct. After seeing the MIT Whirlwind project on a visit to America in August 1952, he realised that EDSAC 2 must use the same ferrite core memory. The only problem was that he needed to raise a further £10,000 to pay for it, but fortunately the Nuffield Foundation understood his arguments and put up the money. Wilkes then commissioned Mullard's factory, in Blackburn, Lancashire, which created ferrites, to build the cores to his specification.

David Wheeler, who had left in 1951 to lecture on programming methods at the University of Illinois, returned to Cambridge in September 1953 and proved an immediate help. He even created a 'control matrix', using the cores and his exceptional programming skills, to make using the EDSAC 1 easier. With the control matrix attached, the upgraded computer became known as EDSAC 1.5.

By early 1958, EDSAC 2 was ready to take over computing duties within Cambridge. Keen to reclaim the space, and in a decision Wilkes later regretted, the team disassembled EDSAC 1 that summer and sold most of it for scrap.[21]

And perhaps that is how the EDSAC should be remembered: as a scrappy computer. It wasn't the most beautiful, it wasn't the most reliable, but it worked. And its influence is still felt at Cambridge University to this very day.

---

[21] Fortunately several parts remain intact, and are safely stored at the Science Museum in London. A partial replica is also on show at the UK's National Museum of Computing, found on the Bletchley Park estate, Buckinghamshire.

A complete BINAC system; Albert Auerbach, who ran the first test routine on it, is seated at the rear

Image: courtesy of the Computer History Museum, CC BY-NC-SA

EDVAC, UNIVAC, & Princeton IAS

The computer that never was, and its offspring

While wars are never to be celebrated, they have an undeniable uniting effect. There is nothing quite like an existential threat to your country, to your way of life, when it comes to motivation. This can be seen through the story of the Colossus, where University of Cambridge professors formed an unlikely partnership with the General Post Office, but also in the story of ENIAC. Here, the Moore School of Electronics, part of the University of Pennsylvania, marched in step with the US Army to create the world's first large-scale digital computer.

Then war stops. Organisations return to their natural self-serving state, people look to enrich themselves, personal enmities that were set aside for the good of the country resurface.

This night-and-day difference between peacetime and wartime is highlighted by the different routes taken by the ENIAC and the EDVAC. Both can consider the Moore School as their spiritual birthplace, but while the ENIAC's progress was marked by how well academic institutions and the army worked together – stimulated, admittedly, by large amounts of money – the EDVAC's path went wrong almost from the outset.

Its story is one of ego, of capitalism, of something approaching chaos. Of an idea born of war that could only survive peace by dividing into three rival projects.

But John von Neumann knew none of this on the fateful day he bumped into Herman Goldstine on a railroad platform in the summer of 1944. As discussed in Chapter 6 on ENIAC, this wasn't a complete coincidence. Both men had good reason to be visiting the Army's Ballistic Research Laboratory (BRL) at its Aberdeen Proving Grounds: von Neumann in his role as a consulting scientist for the army; Goldstine as the key liaison between the US Army, which had commissioned the ENIAC, and the Moore School that was building it.

When their paths crossed, von Neumann had been scouring the country for high-speed computational devices for months. The work being done at Los Alamos, groundbreaking research on atomic and hydrogen bombs known collectively as the Manhattan Project, required colossal amounts of calculations. Calculations that weren't feasible on existing equipment, so his task was to hunt out quicker machines. He had already visited Harvard to see the Harvard Mark I in action, but found it wasn't powerful enough and, besides, already booked up by Navy demands.

His visit to Bell Labs, which had created its own electromagnetic relay-based computers (see Complex Number Calculator, Chapter 3), was far more positive. "I spent the better part of a day with [George] Stibitz," von Neumann wrote in April 1944,[1] "who explained to me in detail the principles and the working of his relay counting mechanisms, and showed me the interpolator as well as the almost-finished anti-aircraft fire-control calculator ... Dr Stibitz even suggested, what went far beyond my expectations, that he may seek permission for an experimental computation of the kind I suggested on the big machine in the process of breaking it in".

Four months later, he was still keen. In a letter to Robert Oppenheimer, director of the Los Alamos Laboratory, he wrote that Stibitz's big machine "would be well worth having if a future of, say 1½ years or more is being envisaged for the project".[2] By this time he had also met Leland Cunningham, who oversaw the machine-computing section at BRL in Aberdeen. "In Cunningham's opinion the simplicity in planning, the reliability of the elements, the self-checking features, and the ability to run overnight without a crew, should alone make the [Stibitz] machine five times or more faster than any IBM aggregate, quite apart from the other advantages."

The fact that nothing came of this stems from von Neumann's chance meeting with Goldstine. One that happened – historians believe[3] – soon after von Neumann sent the above letter to Oppenheimer. It was a story Goldstine told many times, which we cover in Chapter 6 on ENIAC, but it bears repeating. "Along came von Neumann with a very abstracted look on his face and he and I are the only two people on the railroad platform," he recalled in 1967.[4]

Goldstine says that he considered leaving the great mathematician alone, but "my egotism got the better of me". Von Neumann, who in all accounts is described as approachable, even funny, was happy to chat. But the mood of the conversation shifted rapidly from "polite chit-chat" to a grilling when Goldstine explained his involvement with the ENIAC. "And he immediately got exceedingly interested. I didn't at that time realise why but he spent the rest of the time of that day and several days subsequent to that in quizzing me about this machine."

---

[1] Letter from von Neumann to Warren Weaver, 10 April 1944, quoted in William Aspray, *John von Neumann and the Origins of Modern Computing* (MIT Press, ISBN 978-0262518635), p32
[2] Letter from von Neumann to Robert Oppenheimer, 1 August 1944, quoted in William Aspray, *John von Neumann and the Origins of Modern Computing*, p33
[3] It's a cause of some frustration to historians that there is no written record of the date, with Goldstine's recollections tending to use woolly phrases such as "sometime in the summer" in his retellings.
[4] Association for Computing Machinery Meeting, 30 August 1967, Archives Center, National Museum of American History, tape 1

At some point in early August – again, Goldstine is hazy on dates – von Neumann visited the Moore School for the first time. "I recall with amusement Eckert's reaction to the impending visit," wrote Goldstine.[5] "He said that he could tell whether von Neumann was really a genius by his first question. If this was about the logical structure of the machine, he would believe in von Neumann, otherwise not. Of course, this *was* von Neumann's first query."

This marked the beginning of von Neumann's many visits to the Moore School, and ultimately the ENIAC would serve the purpose he sought: in December 1945, into early 1946, it was used to calculate thermonuclear equations crucial to the development of the hydrogen bomb. In terms of the EDVAC's story, what matters is that while von Neumann arrived too late to have any notable impact on the ENIAC's design, with building well under way, he would go on to have many discussions with the Moore School team, including Goldstine, on the design of its successor.[6]

Much of this time is now shrouded in mystery, in part due to a lack of detailed written records but also because the various actors – most notably Presper Eckert, John Mauchly, Goldstine, and von Neumann – came to have entrenched positions depending on where they stood on the patents that emerged from the EDVAC's creation. So here we will focus on facts cemented by contemporary documentation; any retrospective quotes from the people involved should be sprinkled with generous helpings of salt.

We know from a three-page memo[7] that in late January 1944, Eckert had considered a "Magnetic Calculating Machine" that replaced many of the costly valves used in the ENIAC with a magnetic "disc or drum" for storage. He also specified that it should use binary, but there was little space in this document for detail on how calculations might be carried out. It is an overview, nothing more.

We also know that on 11 August, Goldstine wrote to Colonel Leslie Simon suggesting that the US Army should grant a second contract to the Moore School "to permit that institution to continue research and development with the object

---

[5] Herman H Goldstine, *The Computer from Pascal to von Neumann* (Princeton University Press, 1972, paperback edition 1993, ISBN 978-0691023670), p182
[6] Like most of these early computers, the ENIAC was frequently upgraded during its long life, and von Neumann assisted in this process in the late 1940s. So while he didn't have an impact on the first version of the ENIAC, he is considered to have had an active part in its later development.
[7] Presper Eckert, 'Disclosure of Magnetic Calculating Machine', 29 January 1944, **rpimag.co/eckertdisclosure**

Robert Oppenheimer and John von Neumann in front of the Princeton IAS computer, 1952
Image: Public Domain

of building ultimately a new ENIAC of improved design".[8] In particular, the new computer would have more storage, fewer valves, and greater capacity in terms of the number of problems it could solve. It would also be truly programmable: reprogramming the ENIAC involved rewiring it using plugboards. A process that could take weeks, depending on the program's complexity.

At some point in August, Goldstine recollects, Eckert "came up with the idea that a mercury delay line could be used for storage of information".[9] This was one of his greatest contributions to early computers, with many of the computers in this book relying on this fussy but ingenious technology. The idea came to him because he had previously worked with mercury delay lines as part of a radar project.[10]

And August 1944 wasn't done yet. Spurred on by Goldstine's earlier memo and perhaps with the added impetus of von Neumann's interest, on the 29th of that month the Firing Table Reviewing Board convened to consider their next move. Attendees included Cunningham and Goldstine, plus five others from the BRL, along

---

[8] Memo from Goldstine to Simon, 'Further Research and Development on ENIAC', 11 August 1944, reprinted in Herman H Goldstine, *The Computer from Pascal to von Neumann*, p185
[9] As above, p186
[10] William Aspray, *John von Neumann and the Origins of Modern Computing*, p36

with two mathematical consultants and a certain John von Neumann. We don't know what role von Neumann played in the decision, but he must have been pleased with the result: the board recommended entering a contract with the Moore School for a new electronic computer.

One of the few things everyone agrees on is that von Neumann regularly visited the Moore School over the following months to discuss the EDVAC – as it quickly became known, standing for Electronic Discrete Variable Automatic Computer – with Eckert, Mauchly, Goldstine, Arthur Burks, and other engineers involved with the project. And on 31 March 1945, everyone seemed to be getting along just fine, if the project's first summary report is to be believed.

Aside from "problems of logical control", the report states,[11] they had informally discussed "the use of EDVAC, storage capacity, computing speed, sorting speed, the coding of problems, and circuit design". To a lesser extent, they covered input and output systems. But here's the most fateful sentence: "Dr von Neumann plans to submit within the next few weeks a summary of these analyses of the logical control of the EDVAC together with examples showing how certain problems can be set up."

Which he did. In April, he mailed his handwritten notes to Goldstine, who then organised for them to be typed up, copied, bound, and distributed to a select group of 31 people.[12] These included engineers working on the project, the army committee that had commissioned the new computer, and a handful of trusted experts such as Douglas Hartree. It was called a 'First Draft Report on the EDVAC' because that's what it was: there were omissions and mistakes. In places, it's confusing. At the same time, for the first time ever, it set out a logical structure for a computer that we would now recognise.

Another notable point about the report was that it only had John von Neumann's name on it. This was problematic. While the logical interpretation is purely von Neumann's, Mauchly and Eckert claimed that it was built off the back of their ideas, which they had shared freely within the group discussions described above. The report made it look like the work was von Neumann's alone.

This wasn't merely a blow to Eckert and Mauchly's egos, and their place in history, but also threatened any future applications for patents. They attempted to

---

[11] William Aspray, *John von Neumann and the Origins of Modern Computing*, p38. The report's authors are listed as Eckert, Mauchly, and S Reid Warren.
[12] Thomas Haigh, Mark Priestley, and Crispin Rope provide a detailed timeline in their comprehensive tome, *ENIAC in Action* (The MIT Press, 2016, ISBN 978-0262033985), pp137-139

redress the potential damage – or, in their view, set the record straight – by publishing their own report entitled 'Automatic High Speed Computing: A Progress Report on the EDVAC' in late September.[13]

The most pertinent aspects of this report come at the start, as part of the 'Historical Comments' section. After discussing the creation of the ENIAC they describe the need for a new machine, the EDVAC. "It was clear that this new machine would, with much less equipment, easily handle problems beyond the intended scope of the ENIAC. Therefore, by July, 1944 [so before von Neumann met them] it was agreed that when work on the ENIAC permitted, the development and construction of such a machine should be undertaken."

The EDVAC installed at the Ballistic Research Laboratory
US Army photo, Public Domain

And what of von Neumann's role? "He has contributed to many discussions on the logical circuits of the EDVAC, has proposed certain instruction codes, and has tested these proposed systems by writing out the coded instructions for specific problems. Dr von Neumann has also written a preliminary report in which most of the results of earlier discussions are summarised. In his report, the physical structures and devices proposed by Eckert and Mauchly are replaced by idealised elements to avoid raising engineering problems which might distract attention from the logical considerations under discussion."

In other words, the ideas were ours, not his. Step away from our patents.

---

[13] John Mauchly and Presper Eckert, 'Automatic High Speed Computer: A Progress Report', typed manuscript, 30 September 1945. An incomplete copy can be viewed at **rpimag.co/edvacreport**

Goldstine was very much in the von Neumann camp. In his view, before the eminent mathematician arrived, "the group at the Moore School concentrated primarily on the *technological* problems, which were very great; after his arrival he took over leadership on the *logical* problems". (The italics are Goldstine's.) To a large extent, this view is reflected in the differing emphases of the two reports, but we know that von Neumann was interested in the practical side of computer development too. For example, in his letters to Goldstine during 1945 he mentions cathodes and voltages; he is well aware of the electrical engineering side of creating a computer.

One anecdote, shared by Arthur Burks,[14] highlights the fight between clean logic and real-world electronics. "I remembered well a discussion of serial adders that took place at one of our meetings [at the Moore School] of March 1945. Pres [Eckert] and John [Mauchly] had designed several serial adders, the simplest of which took ten tubes. Not knowing of these results, von Neumann announced cheerily that he could build an adder with five tubes. We all looked amazed, and Pres said, 'No, it takes at least ten tubes.' Johnny [von Neumann] said, 'I'll prove it to you,' rushed to the board, and drew his adder.

"'No,' we said, 'your first tube can't drive its load in 1 μsec, so an inverter is needed, then another tube to restore the polarity.' And so the argument went. Johnny was finally convinced. But he was not taken aback. 'You are right,' he said. 'It takes ten tubes to add – five tubes for logic, and five tubes for electronics!'"

This story also serves as a reminder to the quiet role played by Arthur Burks in the EDVAC's creation. He had a Philosophy PhD and background in logic, but had earned his engineering spurs the hard way through his work on the ENIAC. His part in the computers' development is often underplayed, but Irven Travis, who played a pivotal role in the Moore School EDVAC's story, was unstinting in his praise: "In terms of logical design and mathematical intuition, if you will, Arthur Burks was one of the most brilliant men we've had in that field – ever," he told historian Nancy Stern.[15] He was also often part of the meetings where the EDVAC's structure took shape over late 1944 and early 1945.

---

[14] Arthur Burks, 'From ENIAC to the Stored-Program Computer', reprinted in N Metropolis, J Howlett, and Gian-Carlo Rota (eds.), *A History of Computing in the Twentieth Century* (Academic Press, 1980, ISBN 978-0124916500), p341
[15] 'Oral history interview with Irven Travis' by Nancy Stern, 21 October 1977, Charles Babbage Institute, University of Minnesota, **rpimag.co/irventravisinterview**, p16

Presper Eckert demonstrating a BINAC memory unit, c.1948-9
Image: courtesy of the Computer History Museum, CC BY-NC-SA

Von Neumann summarised the difficulty of ascribing credit during a fraught meeting to discuss patents in 1947: "There are certain items which are clearly one man's," he is quoted as saying in the minutes, "[such as] the application of the acoustic tank to this problem was an idea we heard from Pres Eckert. There are other ideas where the situation was confused. So confused that the man who originated the idea had himself talked out of it and changed his mind two or three times. Many times the man who had the idea first may not be the proponent of it. In these cases it would be practically impossible to settle its apostle."[16]

To an extent, the argument was academic: the lightning was out of the bottle. On his return to England, Douglas Hartree shared his copy of the report with Maurice Wilkes, who read it avidly as we describe in the story of the EDSAC (Chapter 8). A

---

[16] Remarks from Minutes of Conference held at the Moore School of Electrical Engineering on 8 April 1947 to discuss patent matters, reprinted in Herman H Goldstine, *The Computer from Pascal to von Neumann*, p195

copy also found its way to Alan Turing, who referred to the report when setting out his plan for the ACE.

Perhaps the biggest irony is that the EDVAC, as laid out logically by von Neumann and physically by Eckert and Mauchly, never came into being. From this point, it would split into three rival projects. The first, for the US Army, would be built at the Moore School as per the contract. While the eventual computer would bear the name EDVAC, it was substantially different from these mid-1945 plans.[17] The second line of development would follow Eckert and Mauchly, who as we shall soon see left the Moore School to set up their own company. And the third would be built by von Neumann and friends, the Princeton IAS computer.

The split came gradually, grumbling tremors that acted as precursors to an earthquake. It's surely no coincidence that these first tremors came at the point where the war in Europe had ended and the USA felt that victory in the Pacific was inevitable: everyone was looking at what was to come next. For von Neumann, a return to his mathematical research; for Eckert and Mauchly, perhaps their chance to make a fortune?

By now, von Neumann was hooked on high-speed electronic computers: he wanted to build one dedicated to research, much like his friend and fellow mathematician Max Newman on the other side of the Atlantic (see the story of the Pilot ACE, Chapter 10). As early as February 1945, von Neumann discussed the make-up of a team who could build such a computer with Goldstine and Cunningham – his wish list of the time included Eckert, Mauchly, and Stibitz.[18]

In March, Norbert Wiener from MIT approached him with a tempting proposition: join as chair of mathematics and he would have a ready-made lab at his beck and call. This was in stark contrast to Princeton's IAS (Institute for Advanced Study) which, at that time, purely focused on theoretical research. MIT was in many ways the logical choice for a new computer, home as it was to Vannevar Bush's differential analyser and a haven for applied mathematicians. Von Neumann visited in August 1945 and was given a formal job offer the following month. But he didn't take it.

---

[17] This is set out brilliantly in a paper by Michael Godfrey and DF Hendry called 'The Computer as von Neumann Planned It'. It was published in the *IEEE Annals of the History of Computing*, Vol 15, No.1, 1993, pp11-21, and a copy can be downloaded from **rpimag.co/computervonneumann**

[18] Nancy Stern, *From ENIAC to UNIVAC* (Digital Press, 1981, ISBN 978-0133315059), p264, note 47

That's partly because the IAS didn't want to lose such an influential figure from their staff, but also because von Neumann felt loyalty to the institute: it had offered him a position long before his reputation was made. Not that von Neumann was so loyal that he kept quiet about his openness to a move when visiting rival institutions. Following such an occasion, the University of Chicago invited von Neumann to create an Institute of Applied Mathematics, while he received what historian William Aspray describes as "job feelers" from Columbia and Harvard.[19]

James Pomerene working on the IAS machine, holding a Williams tube
Image: Wikimedia Commons, CC BY-SA 4.0

If this was von Neumann employing the time-honoured tactic of using other job offers as leverage, it worked. Despite objections from the more conservative trustees, the IAS director, Frank Aydelotte, added his support to the IAS computer project and von Neumann politely turned down his suitors.

In the meantime, he had been hard at work finding financial backers to build a computer in partnership with the IAS. The main thrust of his argument, particularly to potential military partners already contributing to large-scale computing projects elsewhere, was the need for a dedicated scientific computer. "[It] seems to me that it would be an essentially incomplete policy to develop such devices only for industrial or government laboratories, which have definite, and necessarily relatively narrowly defined, applied problems to which they must devote all or most of the time of their equipment," he explained in a letter to Commander Lewis Strauss of the US Navy.[20]

---

[19] William Aspray, *John von Neumann and the Origins of Modern Computing*, p51
[20] Miklós Rédei, *John von Neumann: Selected Letters* (American Mathematical Society, 2005, ISBN 978-0821837764), letter from von Neumann to Strauss dated 20 October 1945, p236

He concluded: "I have no doubt whatever that we are here on the threshold of very important developments both in pure mathematics and in its applications, and that a pure research institution should spend several years in building a machine and experimenting with it. If we devote in this manner several years to experimentation with such a machine, without a need for immediate applications, we shall be much better off at the end of that period in every respect, including the applications."

His argument paid off. On 6 November 1945 the Electronic Computer Project came into being with joint funding from the IAS, Princeton University, the Office of Navy Research, and one private partner: the Radio Corporation of America, better known as RCA.

To partner with the RCA made sense on two fronts. First, much like Bell Labs, it had built computing devices for the military during the war. Second, von Neumann already knew two key figures there: Vladimir Zworykin, who invented an early cathode ray tube called the ionoscope, and Jay Rajchman who was working on a CRT storage technology called Selectron. The RCA funding would cover the cost of the storage for the computer, but they would keep the patents.

With funding in place, von Neumann started to build his team. His first recruit was Goldstine. Von Neumann offered him the role of deputy director for the computing project in late November. Goldstine took it up once the ENIAC was unveiled in February 1946. Arthur Burks followed a month later, but on a temporary basis: having decided to return to his first love of philosophy on a full-time basis, he had already accepted a position at the University of Michigan. But before the new academic year started, he agreed to work with Goldstine and von Neumann on the computer's logical design.

They also needed engineering talent, and despite their differences of opinion over patents von Neumann still wanted Presper Eckert to head up the team. He offered Eckert the job on 27 November 1945, but Eckert took time weighing up his options. "Eckert was being torn by Mauchly, who of course wanted to stay associated with Eckert, because Mauchly really needed Eckert," said Goldstine.[21] "He was torn by his wife, who wanted to stay in Philadelphia close to her family. He was torn by his

---

[21] 'Oral history interview with Herman Goldstine' by Nancy Stern, 14 March 1977, Niels Bohr Library & Archives, **rpimag.co/goldstineinterview**, p37

parents, who wanted him to stay in Philadelphia. And he was torn by his own desire to make a lot of money."

But von Neumann didn't help matters in January 1946. Francis Reichelderfer, Chief of the US Weather Bureau, wanted to know how electronic computers might help advance meteorology, and asked the RCA's Zworykin to meet with his representatives.[22] Zworykin invited von Neumann along, and they met on 9 January to discuss the exciting new opportunities that lay ahead. This meeting wasn't the problem: that came the next day, when *The New York Times* published an article entitled 'Electronics to Aid Weather Forecasting'.

The story told of "a new electronic calculator, reported to have astounding potentialities, which, in time, might have a revolutionary effect in solving the mysteries of long-range weather forecasting". It went on to name von Neumann and Zworykin as the inventors of this new machine, with the clear implication to any readers being that the RCA and von Neumann were the big forces in this new space.

Von Neumann quickly apologised to the Moore School team and the US Army, who had also been annoyed by the coverage: the ENIAC's official unveiling was due the following month, so this was terrible timing. Eckert was still fuming almost 30 years later. "You know, we finally regarded von Neumann as a huckster of other people's ideas with Goldstine as his principal mission salesman," he told Nancy Stern in 1977,[23] in relation to the First Draft report. He soon added: "Von Neumann was stealing ideas and trying to pretend work done at the Moore School was work he had done." Stern asked, "Over matters like *The New York Times* leak and things like that, I assume you mean?" A terse, one-word reply: "Yes".

The only surprise, then, is that it was von Neumann who withdrew the offer – in March 1946 – before Eckert refused it. By this time, the ENIAC had been unveiled to the world, and the Moore School realised it could be sitting on a money-making pot of patents. And this would lead to one final tremor, rising to a team-destroying earthquake, that resulted in Eckert and Mauchly both abruptly leaving the Moore School.

---

[22] Nancy Stern, *From ENIAC to UNIVAC*, p83
[23] 'Oral history interview with J Presper Eckert' by Nancy Stern, 28 October 1977, Charles Babbage Institute, **rpimag.co/eckertinterview**

It came in the form of Irven Travis, who had been an assistant professor at the Moore School before being drafted to the US Navy. He'd spent the final four years of the war in the Bureau of Ordnance in Washington DC in charge of the contracts with organisations such as Bell Labs, RCA, MIT, and the Moore School. On his return in January 1946, he was the obvious person to take up a role as director of research for the organisation.

Right from the start, Travis knew that one of his biggest priorities was to sort out patents. "The university was in an untenable position," he said in 1977.[24] "It had an obligation under a contract with United States Government that it couldn't fulfil. It had no contract with any employees. So one of the first things I did was to write up a patent agreement, more or less along the lines that I had known about at MIT and Bell Labs."

This agreement boiled down to signing away any past and future patent rights to the university, or leaving with immediate effect. Most of the Moore School engineers chose to sign, but it forced Eckert and Mauchly – the two men pursuing patents for their work on the computer projects – to leave at the end of March. Eckert, as he was wont to do, put it bluntly: "I have a letter from Pender in the safety deposit box which says I resigned; but I was fired, by Irv Travis."[25] The Pender here refers to Harold Pender, dean of the Moore School.

Burks, who had already decided to leave, took a more pragmatic view. "I think that's too extreme to say that they were forced out," he said to Nancy Stern in a 1980 interview, when she put forward Eckert's point of view. "[Every] engineer was told 'if you want to stay in this role you will have to agree to abide by this long-standing, but previously unenforced, university policy on patents'."[26]

Both Eckert and Mauchly had had an easy option, especially if they wished to stay working together, as they had both been approached by IBM about setting up a computing department there, and they also knew they had built enough of a reputation to start their own company. Eckert was in favour of playing it safe and taking the IBM job, but a combination of Mauchly and Eckert's first wife (who didn't think he would be happy at IBM) meant that they decided to go their own way.

---

[24] 'Oral history interview with Irven Travis' by Nancy Stern, 21 October 1977, Charles Babbage Institute, University of Minnesota, **rpimag.co/irventravisinterview**, p29
[25] As above, p52
[26] 'Oral history interview with Arthur W and Alice R Burks' by Nancy Stern, 20 June 1980, Charles Babbage Institute, University of Minnesota, **rpimag.co/burksinterview**, p69

Eckert, however, always had his reservations. "I didn't think we'd have enough money for this development [of new computers] and as it turned out I was right," he told Stern. "I also think that it was probably a mistake [to go it alone] and it wasn't a mistake. It's healthier for the United States to have a UNIVAC." In 1980, at the time of the interview, UNIVAC computers were still selling in volume and giving IBM healthy competition. Even today, it echoes in the name of global services company Unisys.[27]

The pair's first contract came, indirectly, through the Census Bureau. Eckert and Mauchly had visited them several times over the previous months, but on behalf of the Moore School. The bureau were interested in the idea of an EDVAC-style digital computer, but couldn't commission it directly themselves – they were only allowed to order finished products, not those in development – and so the contract went via the Army Ordnance Department. But even with funds approved, things weren't straightforward: as always with such contracts, the army used an independent expert to assess the proposal. In this case, George Stibitz from Bell Labs.

His response was lukewarm. "There are so many things undecided [about the proposal] that I do not think a contract should be let for the whole job," he wrote to John Curtiss, assistant director of the National Bureau of Standards (which was itself helping the Census Bureau assess the proposed computer's feasibility, such were the layers within layers of government contracts). But Stibitz offered the lifeline that "their suggestion seems promising enough to let a contract to study the problem, leading to a solid proposal and schematic".[28]

Curtiss decided to largely follow Stibitz's advice, deciding to award the young-ish entrepreneurs – Mauchly was 39, Eckert a fresh-faced 27 – an initial "study contract" of $75,000 in June 1946. This was meant to cover the creation of a scale model computer with two mercury delay lines "complete with associated pulse shaping and regenerative circuits" plus a tape transport mechanism.

The contract was for an "EDVAC-type machine", with the name of the new computer formally becoming UNIVAC, standing for Universal Automatic Computer.

---

[27] Unisys was the name given to the organisation formed when the Burroughs Corporation and the Sperry Corporation merged in 1986, with Sperry having previously been Sperry Rand. The official reason given for the name Unisys is that it shortens the words 'united, information, and systems', and was suggested by Christian Lee Machen. But we know the real reason.

[28] Nancy Stern, *From ENIAC to UNIVAC*, p105

in May the following year.[29] On successfully completing this phase, the Census Bureau would award a second fixed fee of $169,600 to complete the full computer. That brought the total to roughly half the minimum Eckert and Mauchly estimated the first UNIVAC would cost to build: between $413,000 and $671,000.[30] Their plan, 'hope' may be a better word, was to absorb the initial loss in the belief that more orders would follow. Plus, they would hold all patents stemming from the work.

With $75,000 in the bank – although the contract only officially came into effect in October – Mauchly and Eckert created their business, initially titled the Electronic Control Company. They hired two floors of a building in downtown Philadelphia and started hiring engineers and programmers.

In today's money, $75,000 translates to roughly $1.25 million. Without more contracts, the company was going to run out of funds long before they built the proof-of-concept computer, so clearly they needed to find money from elsewhere. The answer, of a kind, came from Northrop Aircraft. It was developing a long-range guided missile for the US Air Force and needed a computer to guide its in-flight navigation.

Northrop ultimately wanted "a compact, airborne computer", with one of Northrop's stipulations being that the computer must be compact enough (less than 20 cubic feet) to fit through bomb bay doors.

The price Northrop was willing to pay for this ambitious machine? $100,000, with $80,000 upfront and $20,000 on delivery. It would ultimately cost almost $300,000 to build,[31] and there's every chance that Eckert and Mauchly suspected as much when they agreed the deal in October 1947. But it gave them an injection of funds – and, they hoped, could also act as completion of the study for the Census Bureau, releasing the much-needed balance of $169,600. Whether they believed they could complete the system by the stated end date of 15 May 1948 is a different question altogether.

If you're beginning to have doubts over Eckert and Mauchly's business acumen, you're not alone. Isaac Auerbach was the Electronic Control Company's first employee who hadn't worked on the ENIAC, joining the engineering team right at the start. "Neither Eckert or Mauchly in my opinion were competent managers, competent

---

[29] All quotes in this paragraph are from p106 of *From ENIAC to UNIVAC* by Nancy Stern, p106
[30] Arthur L Norbert, 'New Engineering Companies and the Evolution of The United States Computer Industry', in *Business and Economic History*, Vol 22, No 1, Fall 1993, Business History Conference, ISSN 0849-6825, p186
[31] To be precise, $278,000 according to Nancy Stern, *From ENIAC to UNIVAC*, p123

US Census Bureau employees tabulate data using one of the agency's UNIVAC computers, c.1960
Image: US Census Bureau, Public Domain

leaders, or competent executives, or understood business at all," he said.[32] "They were visionaries, and they were brilliant technically, and they would not let somebody else run the side of the company in which they were inept."

The contract with Northrop, which would eventually produce the BINAC (short for Binary Automatic Computer), is compelling proof for Auerbach's argument. On one hand, it was a triumph: the first stored-program electronic digital computer to operate in the USA. On the other, a loss-making failure, as it never performed to the satisfaction of Northrop.

---

[32] 'Oral history interview with Isaac L Auerbach' by Bruce H Bruemmer, 2-3 October 1992, Charles Babbage Institute, University of Minnesota, **rpimag.co/auerbachinterview**, p5

This wasn't the only ambitious part of BINAC. It was actually two computers in one, with calculations being simultaneously run through two separate arithmetic units: that way, if there was a difference in the answers something had gone wrong. That meant two separate storage systems, each with a 512-word capacity, plus separate power supplies. The only thing they shared was a decimal to binary converter and their input/output mechanism, which could be either electronic typewriter or magnetic tape.

The latter was another first for a computer and an important one in historical terms: magnetic tape remains important as a form of data storage even today (most often as a backup medium). As is often the case with first attempts, though, it did not work well, and would be eventually replaced by "metallic tape plated with nickel", according to historian Nancy Stern.[33]

The BINAC was a small computer by ENIAC standards. Each processor included 700 valves, giving a total of 1400 versus 18,000 for the ENIAC, and stood "five feet high, four feet long, and one foot wide" according to the press release[34] that celebrated its official arrival on 22 August 1949.

This was also the day Northrop formally accepted the machine, more than a year later than the contract stated. So not only had Eckert and Mauchly grossly underpriced the BINAC, they had also underestimated how long it would take to build. Then again, this is hardly unusual: even today, IT projects are legendary for costing far more and taking far longer than originally estimated.

Despite its tardiness, the BINAC scored an important first by being the first stored-program electronic computer to run successfully in the United States. On 7 February 1949, Albert Auerbach[35] ran a test routine on the computer,[36] following that up with a fuller 23-line program to compute squares in March.

It isn't mentioned often enough, but much of the programming work for the early computers – including the BINAC and UNIVAC – was done by women. Betty Holberton was one of the first Electronic Control Computer employees, even working for free at weekends before Eckert and Mauchly had money to pay her.

---

[33] Nancy Stern, *From ENIAC to UNIVAC*, p120
[34] Press release as reprinted in a collection of papers ('The Albert A Auerbach Collection'), historyofscience.com/pdf/BINAC-Collection.pdf
[35] No relation to Isaac Auerbach
[36] Herman Lukoff, *From Dits to Bits: A Personal History of the Electronic Computer* (Robotics Press, 1979, ISBN 978-0896610026 – the 'Dits' refers to his early passion of transmitting the dits and dahs of Morse code), p84

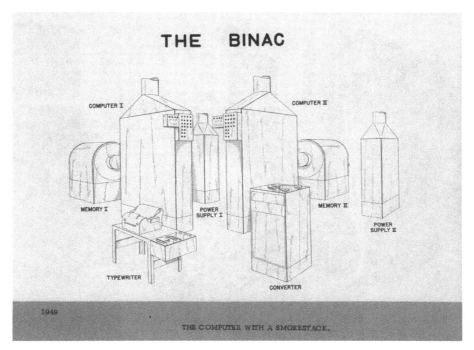

A line drawing of a complete BINAC system, 1949
Image: courtesy of the Computer History Museum, CC BY-NC-SA

That was in early 1946, before the money from the first government contract arrived. Her enthusiasm hadn't dimmed by the time the BINAC was created: "I didn't leave the company. I was on the machine 16 hours and eight hours off and I slept in the ladies' room," for two straight weeks, she said.[37] "I was working on, essentially, the simulation of the sorting system that was to be used on the UNIVAC I in calculating [tape buffer delays]."

Eckert and Mauchly had always hoped that by being the first stored-program computer in the USA they would attract more business, and the press release wasn't afraid to shout of its achievements. "The new computer showed its mettle by solving 'Poisson's Equation,' a typical engineering problem," it stated.[38] "BINAC spent more

---

[37] 'Oral history with Jean J Bartik and Frances E (Betty) Snyder Holberton' by Henry S Tropp, 27 April 1973, Smithsonian National Museum of American History, **rpimag.co/bartikholbertoninterview**, p110

[38] Press release as reprinted in a collection of papers ('The Albert A. Auerbach Collection'), **historyofscience.com/pdf/BINAC-Collection.pdf**

The BINAC computer worked best on sunny days, according to Florence Anderson
Image: courtesy of the Computer History Museum, CC BY-NC-SA

than two hours of actual computation to obtain 26 solutions. For each solution the computer did 500,000 additions, 200,000 multiplications, and 300,000 transfers of control, all in the space of 5 minutes. A man with an adding machine would have [needed] years to complete the same job."

The computer was ready for shipping to Northrop, a journey of several hundred miles by train. As per the terms of the contract, Northrop had responsibility for packing it up securely, and this is where – perhaps – things started to go wrong. Whether the quality of construction was too low, it was packaged poorly or a combination of the two, it took several months of work before the Northrop engineers would obtain usable results from their new computer.

Rumours that it never worked at all were contradicted by two former Northrop employees, who wrote in to the *IEEE Annals of the History of Computing* in reaction to an article by Nancy Stern. Florence Anderson pointed out that at that early "stage of computer development ... the amount of time required for maintenance, the quality of components, and the physical environment were all greatly underestimated. For example, BINAC seemed to operate well on sunny days, but poorly on rainy days".[39] Jerry Mendelson was even more definite: "I can state categorically... the BINAC did run successfully after it was installed in Hawthorne, California. I know this to be a fact because a colleague of mine named Bob Douthitt and I made two absolutely successful uninterrupted runs, each in excess of 25-minute duration."

To the outside world, the BINAC was a success, and it also served as an important landmark in the company's contract with the Census Bureau. Finally, that $169,600 was theirs, even if it didn't cover Eckert and Mauchly's outgoings to date – not to mention the fact they still had to build the UNIVAC itself.

For now, we will leave Eckert, Mauchly, and the rest of the team, and travel back to April 1946 and the Moore School. The brains trust that created the ENIAC and designed the EDVAC – not just Eckert and Mauchly, but also Burks, Goldstine, and von Neumann – have all left. What's left is a contract and the outline of a design. Fortunately for Irven Travis, he not only had a handful of ENIAC old-timers, most notably Kite Sharpless and Carl Chambers,[40] but also Moore School alumni who were returning from the war.

Herman Lukoff, who had just turned 23, was one such young veteran. Travis immediately assigned him to the EDVAC project, giving Lukoff the job of devising a control mechanism for the mercury delay line. Most early mercury-delay lines suffered terribly due to the way their delay properties varied depending on the temperature, with the common solution being to control the temperature. Lukoff took a different approach, varying the gaps between pulses so "that they could be spread or condensed to give the exact number of pulses required to fill the column".[41]

With the help of Steve Chen and Joe Chedaker, over the space of a few months Lukoff solved one of the biggest problems affecting mercury delay line storage: that the timings changed with temperature. Their creations were even ready for public

---

[39] *IEEE Annals of the History of Computing*, Vol 2, No 1, Jan 1980, p83
[40] While Harry Huskey was still working at the university, at this point he was an assistant professor in the Maths department. In June 1946, he was offered the chance to head the EDVAC project and would have taken it, but no-one had consulted with the chair of the department who demanded they withdraw the offer. They did and Huskey resigned. As we've covered in Chapter 10 on the Pilot ACE, he spent a year in Manchester after being invited by Douglas Hartree.
[41] Herman Lukoff, *From Dits to Bits*, p59

demonstration, so with the help of a test assembly and an oscilloscope to show the pulses, Chen and Lukoff headed to New York for the Institute of Radio Engineers show in March 1947.

They set it up the night before the show started and it worked as planned. Then came the public demonstration. "No sooner was the equipment turned on than unusual things started to happen. The mercury memory would suddenly fill up with all kinds of extraneous pulses," Lukoff recalled.[42] The crowd wanted to know what was going wrong, and it was only when he noticed that the pulses were changing "rhythmically every two seconds" that he caught sight of "a rotating Army Signal Corps radar antenna halfway across the hall". Lukoff convinced the operator to switch off the power and the demonstration worked perfectly once more.

Unfortunately, the rest of the Moore School's EDVAC project did not run so smoothly. Without Burks, Mauchly, or von Neumann, they were relearning the fundamentals of computer logic design. Even the gifted Lukoff was struggling to get to grips with the problem. "Many months were spent on the paper design of the logic," he wrote.[43] "Each time we reviewed the drawings, we would invariably find some illogical condition we had not thought of before that required a redesign." Nor, according to Carl Chambers, was von Neumann's First Draft of any use, describing it as coming "from the point of view of a neurologist" and "of not much practical help".[44]

Amidst all this, one amazing, almost miraculous event took place: a special course entitled 'Theory and Techniques for Design of Electronic Digital Computers'. But it became universally known as the Moore School Lectures, and has reached legendary status for not only the wealth of information shared but the influence it went on to have. Over the course of 48 lectures, taking place at the Moore School from 8 July to 30 August 1946, it equipped the students with the information they needed to start building their own computers. Even if there were still many problems they would need to solve themselves – such as building mercury delay lines.

The lecturers read like a Who's Who of computing in America at the time. George Stibitz kicked things off with an introduction to the course where he explained

---

[42] Herman Lukoff, *From Dits to Bits*, pp61-62
[43] As above, p62
[44] Thomas Haigh, Mark Priestley, and Crispin Rope, *ENIAC in Action*, p147

why electronic digital computers were already so important, and were set to become more important in the future. Other early computer luminaries giving talks included Howard Aiken, John von Neumann, Herman Goldstine, Douglas Hartree, and Jan Rajchman. Perhaps surprisingly, given their abrupt departure mere weeks earlier, the bulk of the lectures were given by Mauchly and Eckert. Perhaps it helped that they were both given $1200 each for doing so, and at that point no money had yet arrived from the Census Bureau contract.

Vacuum tubes in a UNIVAC I computer
Image: Diomidis Spinellis, CC BY-SA 4.0

The list of attendees was no less impressive. There were senior people from the National Bureau of Standards, MIT, the Naval Research Laboratory, General Electric, not to mention Maurice Wilkes (who was so inspired by the conference he built the EDSAC) and Claude Shannon of Bell Labs. Officially the total came to 28, with invitations strictly limited, but we know that gate-crashers included Cuthbert Hurd of IBM and the MIT's Jay Forrester.[45]

The main reason for holding the event was that after unveiling the ENIAC to the world, the Moore School had been beset by calls from industry and government departments alike to find out more. This, Pender thought, was the most efficient way to share their learnings. Even better that the army and navy would jointly pay for it.

They recorded the lectures, with this unenviable job shared between Lukoff and his Moore School colleague Dick Merwin. "Wire recordings were not very foolproof; if you think rewinding tape on a reel that has spewed on the floor is a pain, try rewinding fine wire without getting a kink in it," he wrote.[46] The results were

---

[45] Martin Campbell-Kelly and Michael R Williams (eds.), *The Moore School Lectures* (The MIT Press and Tomash Publishers, 1985, ISBN 978-0262031097), pxvii
[46] Herman Lukoff, *From Dits to Bits*, p60

painstakingly transcribed by Moore School secretaries before being turned into a published set of lectures in late 1947, with supplementary material and notes supplied by the lecturers where available.

Unfortunately for the Moore School's EDVAC project, this was a lonely bright spot in a difficult period of transition. Lukoff was eventually lured away by Eckert and Mauchly with the promise of more pay and better prospects, and it's telling that the same day he joined their company – 1 September 1947 – another Moore School colleague joined with him. Kite Sharpless and Carl Chambers had also departed in recent months, leaving the project without a clear technical director until Richard Snyder appeared on the scene. He not only saw the project to its conclusion, but supervised its running in Aberdeen at the BRL.

By this point, they were no longer technically building the EDVAC but the EDVAC 1.5. This name came about after a meeting between the Moore School and the BRL in October 1946 on exactly what they were going to build, which still hadn't been formally defined. The Moore School proposed three possible computers of increasing complexity, with EDVAC 1 a binary computer that could add, subtract and multiply, and with a 1000-word memory. The 1.5 came about because during the meeting they decided to add division in hardware (rather than being programmable) and adding internal checking for all arithmetic operations.

Then, seven months later, they reconvened and decided that a 1.5B version could be created with floating-point operations and an extended instruction set. The floating-point hardware was later dropped as it proved too complicated. Like the BINAC, they also decided that to detect errors they would need to build two identical arithmetic units to check against one another. And now, finally, in May 1947, they finalised the design, almost three years after the computer was first conceived.

Rather than attempt to build it at the Moore School, construction was subcontracted to the Reeves Instrument Corporation. In Michael Williams's 1993 article, 'The Origins, Uses, and Fate of the EDVAC', a footnote explains that the overconfident company thought it could build several copies and promised their first one to the Rand Corporation in return for $100,000.[47] They confidently predicted delivery in May 1948, but when a Rand representative arrived in April in the full

---

[47] Michael R Williams, 'The Origins, Uses, and Fate of the EDVAC', in *IEEE Annals of the History of Computing*, Vol 15, No 1, 1993, p29

expectation of seeing "EDVACs rolling off the production line" he discovered, well, nothing at all. Reeves quickly announced that it was exiting the computer business before it had even started.

It turned out that creating computers in the late 1940s was a lot tougher than first draft reports made it look, so it must have been with some relief that the Moore School – which did eventually build the EDVAC 1.5B for an estimated total cost of $467,000[48] – delivered its machine to the Aberdeen Proving Ground in late 1949. Unlike the BINAC, this was a big computer, consisting of a 2700lb (1.2-tonne) memory unit and eight further units (such as arithmetic units and controllers) weighing 850lbs (385kg). Not to mention a 4000lb (1.8-tonne) power supply.

At the time of delivery, it included around 3563[49] vacuum tubes and 50,000 feet of wiring, measuring 30 feet long by 14 feet wide. That's a little over nine metres and four metres respectively. There wasn't much room left in the room that housed it for people or the various pieces of add-on equipment that appeared over the years. We should also feel sympathy for the operator of the paper-tape input system – a late substitute for a magnetic wire system that never worked properly – who had to pull it by hand past the read head.

But there was more bad news for the Army scientists who wished to operate the EDVAC: it wouldn't work smoothly for years. It ran its first proper program, as opposed to test routines, on 28 October 1951, some two years after it was delivered. "Even then," wrote Michael Williams,[50] "it took a further three months before it was considered reliable enough to run a large calculation – to find the eigenvalues of a 12×12 matrix in January 1952."

Later that year the faster and more reliable ORDVAC, based on the Princeton IAS, was delivered, pushing the EDVAC into third position behind the still operational ENIAC. It was only after that computer blew its metaphorical fuse during an electrical storm in October 1955 that the BRL focused their time and effort into improving the reliability and utility of the EDVAC. Which they did, and for the next five years it turned into something of a workhorse, operating productively for over 130 hours per week on average.

---

[48] As above, p30
[49] The physical measurements and weights come from Michael Williams's article; the number of vacuum tubes from *The Computer as von Neumann Planned It* by Michael Godfrey and DF Hendry.
[50] Michael R Williams, 'The Origins, Uses, and Fate of the EDVAC', p30

As the arrival of the ORDVAC gives away, the Princeton IAS computer had enjoyed a much smoother journey to completion. If the Moore School was a magnet pushing away its most talented staff, the IAS was its polar opposite dragging talented people in. By the middle of 1946 it had hired top engineers such as James Pomerene, Ralph Slutz, Willis Ware and, at the head of them all, Julian Bigelow. Not all of them had computing experience – how could they – but were all highly respected engineers.

Then, on the logic side, the IAS project boasted an all-star team of Burks, Goldstine, and von Neumann. Even before Burks and Goldstine formally set to work in March, von Neumann had thought in great detail about the principles and specifications of a next-generation computer. Keeping in mind the difficulty of the problems he wanted to solve, von Neumann specified the use of binary numbers with 40-digit accuracy and a capacity of 4000 words. He also specified the need for automatic checking facilities. He believed it could all be done with "1000 to 2000 off-the-shelf vacuum tubes" plus "50 to 100 special tubes", according to historian William Aspray.[51]

This was the outline von Neumann presented to Burks and Goldstine when they joined him at the IAS in March 1946. They couldn't have asked for a stronger intellectual head start, and this aura was only added to by the presence of pre-eminent scientists such as Albert Einstein and Kurt Gödel.[52] "Kurt Gödel didn't have a secretary, didn't want one, I assume," said Burks. "So for that summer, when of course we didn't yet have a building for the computer, Herman and I occupied the secretary's office next to Gödel's office. It had a blackboard on the wall. We spent most of our time the first few months planning this new machine, working out the structure and the instructions, and we would consult periodically with von Neumann."

The first fruit of their labours came in the form of a report, published in late June 1946, entitled 'Preliminary Discussion of the Logical Design of an Electronic Computer'. Unlike the First Draft, which presented an abstracted, idealised version of how a computer might work (and was full of errors), this paper goes into prosaic detail. For example, it explains how a computer might use Selectron tubes for storage.

---

[51] William Aspray, *John von Neumann and the Origins of Modern Computing*, p63
[52] To paraphrase his *Encyclopedia Britannica* entry, Austrian-born mathematician, logician and philosopher Kurt Gödel is considered one of the greatest logicians since Aristotle "due to his incompleteness theorem which states that within any axiomatic mathematical system there are propositions that cannot be proved or disproved on the basis of the axioms within that system".

Still, Neumann's love of anthropomorphism remains evident in the section titles of 'The Memory Organ' and 'The Arithmetic Organ'.

One point is worth labouring here. The trio wanted to share their findings, as is common in scientific literature; von Neumann also felt it was only right that, as government agencies were funding a big proportion of the work, there was no place for patent chasing by individuals. 175 copies of the report were printed and freely distributed to interested readers, with the promise of a second part of the report to follow (this would in fact follow as three separate reports).

Burks left at the end of August, as had always been planned, to take up a position as assistant professor of Philosophy at his alma mater, the University of Michigan (he became a full professor in 1954)[53]. But not before leaving 50 pages of notes on how one might program the computer, a subject covered by one of the later reports.

By this time, the engineering team was starting its work, but under Julian Bigelow's diligent direction they didn't rush things. "The period from June 1946 through June 1947 can fairly be described as one of engineering and organisation rather than actual design," he wrote.[54] Having decided to use readily available components whenever they could, they set out to test the options available (although they had to build the test equipment themselves). Another reason for caution is that as soon as the team crystallised around a decision, they would share that with five other groups around the country[55] in the form of engineering drawings. These groups could then build their own versions of the computer to the IAS design. Any mistakes would be multiplied, and at great cost.

The engineers moved into their newly built lab in January 1947, and by mid-1947 they started to design the IAS computer's arithmetic unit. It's worth reading Bigelow's detailed – but not impenetrable – account of this work, which not only lists successes but also the blind alleys they went down and mistakes he personally made. It's not merely a technical delight, but a guide for project managers everywhere.

By spring 1948 they had finally assembled and tested the arithmetic unit to the point that it could be fully demonstrated to von Neumann. Unlike most

---

[53] For an accurate potted biography of Arthur Burks, head to the University of Michigan's own description at **findingaids.lib.umich.edu/catalog/umich-bhl-90185**
[54] Julian Bigelow, 'Computer Development at IAS Princeton', as printed in *A History of Computing in the Twentieth Century*, p294
[55] These were the Los Alamos Laboratory, the University of Illinois, Oak Ridge National Laboratory, Argonne National Laboratory, and the Rand Corporation.

UNIVAC I control station in Museum of Science, Boston, Massachusetts
Image: Daderot, Public Domain

mathematicians, he was a talented mental arithmetician, so enjoyed trying to outpace the arithmetic unit when given increasingly difficult sums. "The first few times he was right, but then we put in more complicated [numbers] with also a 'left end' carry, and it eventually happened that what he called out as the answer disagreed with the result shown by the adder," wrote Bigelow. "Upon rechecking, Johnny found he had made an error, and acknowledged the victory of matter over mind."[56]

But there was a problem: the RCA was struggling to make the Selectron, and this was key to the Princeton IAS design. For the computer to function as planned, it needed to have instant access to data, which was the key advantage of CRT storage technology over mercury delay lines (where you had to wait for each stored bit of data to emerge). The RCA made bold claims that it could store 4096 digits per tube, but no samples were forthcoming. The chances of it delivering 40 such tubes were looking slim to impossible.

---

[56] Julian Bigelow, 'Computer Development at IAS Princeton', p302

The Princeton IAS team needed a backup plan, so – using the small workshop that formed part of their engineering lab – they started work on a magnetic drum that would be far slower than the Selectron but would at least act as a working data store. They also experimented with their own home-built CRT storage mechanism.

Perhaps both plans might ultimately have worked, but fortune smiled upon them in the form of the Williams-Kilburn tube being developed in Manchester. In June, they received a copy of Williams and Kilburn's report, and so Bigelow flew over to see the work for himself while his deputy Jim Pomerene worked on their own replica of the British design. "I can remember [Williams] explaining it to me," wrote Bigelow,[57] "when there was a flash and a puff of smoke and everything went dead, but Williams was unperturbed, turned off the power, and with a handy soldering iron, replaced a few dangling wires and resistors so that everything was working again in a few minutes."

When Pomerene reported that they had a trial tube working in the IAS lab, storing 16 digits, Bigelow headed back. Their challenge now was to not only devise a way to up that number from 16 to 1024, but to do so reliably across 40 CRTs. These were all 'off the shelf' monitors, but not all models worked as the phosphor layers needed to behave perfectly – in terms of electron emission – if they were to act as storage. The engineering team tested hundreds of potential CRTs in search of potential models, and by the summer of 1949 had their 40-strong collection of working storage tubes.

Now for the small matter of building a working computer. There was the physical component, with assembly performed by the workshop team, but also highly technical issues around timing. Only by January 1950 were they ready to start testing a fully assembled unit, with Bigelow describing the year as one of "extreme pressure for the IAS engineering group; our laboratory building was overflowing with applied scientists of all sorts".[58] Using genuine programs from nuclear physicists and meteorologists (the project had a whole section dedicated to meteorology), they honed their machine over the course of that year.

Over the following spring it was pushed into wider use, reaching the point where errors were more likely to come from the programmers than from the machine. And then in the summer of 1951 its biggest test: Los Alamos scientists ran a huge

---

[57] As above, p304
[58] As above, p307

thermonuclear calculation that ran for 24 hours per day and for 60 days (with brief respites for them to check for errors, of which only six were disclosed).

"The engineering group split up into teams and was in full-time attendance and ran diagnostic and test routines a few times per day, but had little else to do," wrote Bigelow.[59] "So it had come alive."

And it wasn't only the Princeton IAS computer that sprang into being. Between 1951 and 1953, the Los Alamos National Laboratory built the MANIAC based on the IAS designs, the University of Illinois the ORDVAC (for the US Army) and the ILLIAC (for its own use), while the Argonne National Lab built the AVIDAC and the ORACLE. Even other countries got in on the action, with Australia, Denmark, Israel, Japan, and Sweden all creating computers closely based on the Princeton IAS computer.

Commercial companies weren't shy of taking advantage either. 1953's 'Defense Calculator', the IBM 701, was based on an adapted IAS design, while the Rand Corporation created the JOHNNIAC – in honour of von Neumann – the following year.

All these machines were notable for their longevity, with some carrying on their work into the mid-1960s. This meant they outlasted both the Princeton IAS computer, decommissioned in 1960, and most sadly of all John von Neumann himself. He died of bone cancer at the age of 53 in 1957, having left an indelible print on the world through his computing achievements – the computers we use today are still based on what is known as the von Neumann architecture – and an incredible body of mathematical work.

The legacy of the computing lab at Princeton IAS proved far shorter-lived. There had always been a tension between the purely academic leanings of the Institute for Advanced Study and the dirtier, applied work of computing. In 1958, the computing laboratory was closed, control of the computer passed to the main university, and in 1960 the IAS computer was donated to the Smithsonian National Museum of American History. Sadly, there are no plans to put it back on display.

We have one final story to complete in the EDVAC's rich and complex history: what happened next at Eckert and Mauchly's company. Lukoff, who left the Moore School for the Electronic Control Company in September 1947, was immediately impressed

---

[59] Julian Bigelow, 'Computer Development at IAS Princeton', p308

by Eckert. "Pres Eckert's genius was very apparent to everyone as he multiplexed from project to project, asking penetrating questions and offering ingenious new approaches," he wrote,[60] adding that "Pres was the engineer's engineer".

Although others, including Isaac Auerbach, looked at this approach slightly differently. "Every time you would design something, [Eckert] had a change," he said.[61] "[You] never could finish the design before he had a different design; and then that one had to be rebuilt and checked and made to work and the tolerances worked out. I mean, it was every day with something new."

Many people also found him a daunting character to work for, in ways both good and bad. He certainly wasn't a natural people person – unlike the jovial John Mauchly – but inspired fierce loyalty. "His intellect is overpowering, particularly for a person that hasn't worked with him and just meets him casually," said Jean Bartik, one of the earliest programmers hired by Eckert and Mauchly.[62]

But, she added, "working with him is very exciting, because even the slightest thing you say that's new, he recognises it as new. And he immediately builds on it and sees how it can be used in many different ways. So that you increase the sense of your own worth working with this kind of person instead of being intimidated."

It's clear, then, that Eckert and Mauchly's business problems had nothing to do with intellect or ingenuity; it was their inability to structure deals that wouldn't leave them hundreds of thousands of dollars out of pocket. Part of this is understandable: they needed funds and they also needed contracts to give other companies confidence in them. After all, the Electronic Control Company was effectively a startup.

Even from that perspective, it's hard to defend the deal they struck with Prudential. On 8 December 1948, ECC agreed to build a UNIVAC by 15 September 1950 for $150,000. Along with converters for the IBM punch cards on which its data processing system relied: one card-to-tape converter, two tape-to-card converters. Plus twelve tape drives, five line printers and two key-to-tape encoders. As Nancy Stern put it: "Even by 1948 standards, $150,000 was a paltry sum for such an array of revolutionary equipment."[63]

---

[60] Herman Lukoff, *From Dits to Bits*, p74
[61] Oral history with Nancy Stern, p9
[62] 'Oral history with Jean J Bartik and Frances E (Betty) Snyder Holberton' by Henry S Tropp, 27 April 1973, Smithsonian National Museum of American History, **rpimag.co/bartikholbertoninterview**, p110
[63] Nancy Stern, *From ENIAC to UNIVAC*, p142

This deal coincided with the Electronic Control Company incorporating as the Eckert-Mauchly Computer Corporation, EMCC. This restructure meant outside companies could invest, and the market research company AC Nielsen made an offer to take a controlling interest in what it saw as an innovative but poorly run company. Eckert and Mauchly rejected the offer, but Nielsen still placed an order for a UNIVAC, six tape drives, six key-to-tape encoders, and a line printer.

The deal again made little financial sense for Eckert and Mauchly at $151,400, especially as the founders realised that they needed $500,000 of working capital or they would be unable to fulfil their current order book.[64] They needed a big surge of investment and they needed it fast, or their odds of survival were becoming slim.

Which is where a betting company came to its aid. On 8 August 1948, Henry Straus, the vice president of American Totalisator became chairman, with the company owning 40% of EMCC's shares. In return, it paid $438,000 and lent EMCC a total of $112,000, with the loan to mature in January 1950.

It's easy to see why Straus backed the young entrepreneurs. An electrical engineer himself, he had invented the electric totalisator but struggled to make money from it. He only earned his fortune after forming a partnership with the American Totalisator Company and, intrigued by this new era of electronic digital computers, recognised the potential of EMCC.

It was a cash injection at exactly the right time for Eckert and Mauchly, not only meaning they could deliver the BINAC, which was in the final testing stages at this point in time, but also invest heavily in more employees and a new office building. It had been a bumpy start, but Eckert and Mauchly were now thriving and their company booming. It was the American dream writ large.

But as quickly as they had been rescued, their world collapsed: in October 1949, Straus's private plane crashed, killing him, the pilot, and two passengers. Without the backing of Straus, the American Totalisator Company had no interest in EMCC but great interest in reclaiming its loans. Which were set to mature in February 1950.

With banks refusing their advances, bankruptcy loomed unless EMCC could find a new backer. IBM was one obvious buyer, but wary of anti-trust litigation. Other interested parties, which included rivals Burroughs and the National Cash Register, couldn't summon up deals in time. This left only one realistic bidder: Remington

---

[64] Nancy Stern, *From ENIAC to UNIVAC*, p144

Rand. It would pay the American Totalisator $438,000 for its shares, plus $100,000 to share among all EMCC employees who held stock.

The benefits to Remington Rand are clear: for a cash investment of $538,000, it instantly became the second biggest player in the nascent market for electronic computers. IBM being the other. And if you're thinking how odd that a typewriter company would wish to do so, it already sold automated office machines so this takeover wasn't entirely out of keeping. Indeed, we know from Lukoff's account that the company attempted to head-hunt him and fellow EMCC engineers the previous year; they had all turned the offer down. "My first thought [on hearing of the takeover] was that Remington Rand had to buy the company because it wasn't successful in hiring any of its engineers," wrote Lukoff.[65]

While Eckert and Mauchly had not become the millionaires they might have aspired to be, there was the comfort of their share of the $100,000 – plus the guarantee of a job for eight years at $18,000 per year. A generous salary in 1950, roughly $250,000 in today's money, and they would also get a minimum annual dividend of $5,000 from profits deriding from the patents.

But they lost ultimate control of their company, becoming a division of Remington Rand. Straus had been a hands-off chairman, but VP in charge of research General Leslie Groves – the military head of the Manhattan Project – had his own ideas. Having realised that the true cost of building each UNIVAC was around half a million dollars, his first challenge was to either renegotiate the existing contracts or to somehow get them cancelled. After threats – renegotiate or else – from Remington Rand's aggressive lawyers, Nielsen and Prudential took the easy route out and agreed to cancel their contracts in return for refunds. A similar tactic against the government contractors proved less successful, with all its computers delivered at the agreed prices.

By summer, the first UNIVAC was entering its final testing stage. Unlike the BINAC, here they had few major problems to fight, but instead many minor problems. Many of these fell to Eckert who used whoever was in earshot to bounce problems and solutions off, although Lukoff said that only the similarly gifted Frazer Welsh was quick enough to respond in kind.

As an example – although this one was solved by Lukoff and his team – there were problems with the custom-made tape handlers, as circuitry within the handler

---

[65] Herman Lukoff, *From Dits to Bits*, p97

added unwanted noise to recordings. The UNIVAC's tape drives were breaking technological ground, so such problems had to be solved without outside help, and in this case meant spending many hours tracking down the cause of each piece of interference. Such work took weeks of effort.

But all this hard work paid off. At the end of March, after months of testing, rewiring, of rewriting programs, the UNIVAC ran without fault for hours. "On that day, March 30, 1951, the UNIVAC I computer, Serial 1, belonged to the Bureau of the Census and was no longer our plaything," wrote Lukoff.[66] "On that day, the computer industry was born."

---

[66] Herman Lukoff, *From Dits to Bits*, p110

Pilot ACE in the Science Museum, London, UK
Image: Antoine Taveneaux, CC BY-SA 3.0

Pilot ACE

Turing's universal machine made real

So much has been written about Alan Turing that he has become a legend as much as a man. Some accounts, wary of facts getting in the way of a good story, give him credit for shortening the war by two years by single-handedly breaking the Nazis' Enigma code. Others, particularly in the UK, paint him as the father of computing. The truth, which sadly must allow for facts, is much more nuanced. Including his tragic early death.

There are things we can say without fear of contradiction or argument. That Alan Turing was a genius and polymath; a man of such insight that he, more than any of his time, foresaw the era of AI in which we currently live; a great and compulsive runner, to the point where he could have represented his country at the Olympics; a gay man living in an intolerant world.

But we might not have known any of this had he skipped a lecture in the spring of 1935.

"I believe it all started because he attended a lecture of mine on foundations of mathematics and logic," said Max Newman,[1] who would go on to become a key figure in Bletchley Park's code-breaking success and a lifelong mentor to Alan Turing. "I think I said in the course of this lecture that what is meant by saying that [a] process is constructive is that it's a purely mechanical machine – and I may even have said, a machine can do it."

If Newman was correct, his lecture set Turing on course to write a paper that laid the foundations for modern-day computers. Most notable of all is that Turing created the idea of a universal computer to disprove the forbiddingly named Entscheidungsproblem. This translates as 'decision problem', a term likely invented by Heinrich Behmann[2] in the early 1920s.

The Entscheidungsproblem summarises a far longer list of aspirations set out by German Professor David Hilbert, whom Behmann assisted, to formalise mathematics.

First came completeness, that every true mathematical statement could be proved within the system; next, consistency, that no contradictions would appear; and last was decidability, that there should be a method to determine if any given mathematical statement is provable or not. Kurt Gödel had brilliantly dealt with the

---

[1] Jack Copeland, 'The Church-Turing Thesis', in *Stanford Encyclopedia of Philosophy*, 8 January 1997 (revised 18 December 2023), **plato.stanford.edu/entries/church-turing/#NewmCambMath**. The quote is from an interview circa 1976.
[2] As above, **plato.stanford.edu/entries/church-turing/#Ents**

first two aspirations, completeness and consistency, via two theorems in 1931,[3] but this left decidability as a prize for gifted young mathematicians such as Turing to grab.

The Entscheidungsproblem came at the right time for the 23-year-old. Alan Turing had flown through the mathematics course at Cambridge with first-class honours[4] and his university college, King's, recognised Turing's brilliance by awarding him a £200 research studentship so that he could stay on and achieve a full fellowship. In 2025, that £200 translates to over £12,000 ($16,000).

Alan Turing in 1951
Image: Elliott & Fry, Public Domain

While brilliant, Turing also took a unique approach to solving problems that often frustrated his seniors. Others stand on the shoulders of giants, but Turing had a habit of recreating the giant from first principles, often with his own obscure notation. He was also plagued by bad luck. He thought he was the first to solve a two-century-old challenge known as the Central Limit Theorem, only to be told that a key breakthrough had been published in 1920.[5] While he had used a different approach to that proof, it did not break new mathematical ground.

History was to repeat itself with the Entscheidungsproblem, with American mathematician Alonzo Church publishing a proof in March 1936. In the months since Newman's lecture the previous year, Turing had developed his proof to the point where he had a draft ready in early 1936 and submitted it to the London Mathematical Society in May 1936. It was finally published in late 1936.[6]

---

[3] Panu Raatikainen, 'Gödel's Incompleteness Theorems', 11 November 2013 (revised 2 April 2020), plato.stanford.edu/entries/goedel-incompleteness/#HisEarRecIncThe
[4] Technically, he was on the list of 'B-star wranglers', the equivalent of a first-class Maths degree from Cambridge at the time. The 'B' is misleading, referring not to a grade but a schedule of subjects.
[5] The proof was by Finnish mathematician Jarl Waldemar Lindeberg.
[6] Some sources say 1937, but it was published in two parts in Proceedings of the London Mathematical Society in two parts across November and December 1936. See footnote 1 on p5 of *The Essential Turing*, edited by Jack Copeland (Oxford University Press, 2004, ISBN 978-0198250807)

Turing's paper, 'On Computable Numbers, with an Application to the Entscheidungsproblem', established him as an important and original mathematician. But more importantly, it fired Turing's interest in what we would now call computing and the science behind it.

In the paper, he describes the key concept of a Turing machine.[7] Imagine, Turing suggested, a device that could scan any square on a long tape. Each scanned square contained a symbol or was empty. The machine would then scan the square and take an action depending on the rules it had been given: do nothing or erase the symbol (if present) or overwrite it with a new symbol. The scanning apparatus would then move to the left or right.

Turing describes instruction tables, which you can think of as programs that tell it what to do in any possible situation. He then showed that any computable mathematical function could be computed by means of an instruction table, much like a program or app will today.

The final ingredient of the Turing machine is the idea of states. So, if the Turing machine is in state A, it will follow a particular instruction (which can tell it to switch state). In state B, it will follow a different instruction. So in state A, the instruction might say replace a star symbol with a question mark, but in state B it's told to replace a star symbol with an exclamation mark.

"He introduced Turing machines and described them in detail," says computer historian and Distinguished Professor Jack Copeland,[8] referring to Turing's paper. "Then he said, now I'm going to prove that there's a universal machine. The way he does that, and he devotes a whole densely packed section to it, is to create the instruction table for a universal machine. So, if you set up the scanner in accordance

---

[7] Say you want to create a binary adding device. And that you want to add five and six. In binary, five is represented by 101, six by 110. Let's now separate those two numbers with the symbol $ (it could be any symbol). We now have seven squares lined up from left to right: 101$110. For this particular Turing machine, we need four different states: let's call them s0 (start), s1 (carry), s2 (write) and s3 (halt). Each of those states comes with instructions. If the machine is in a carry state, then it must do certain things: if it reads a 1, then it must write 0, move left, and stay in the carry state. If it reads a 0, it must write a 1, move left and transition to the start state. If it hits the $ symbol and is in the carry state then it must write a 1; if not, it must simply halt. Does it work? If we use our example of 110$101, the machine begins in the start state, s0, at which point its instructions are to read the rightmost symbol of both numbers. Scanning 1 and 0 means it must write 1, return to the start state and move to the next symbol on the left. This time it scans 0 and 1, but the result is the same: write 1, move to the left. It now reads 1 and 1, so knows it must write 0 but stay in the carry state. It reads $, which tells it to halt after writing 1. And the tape now reads 1011, or 11 (eleven). Otherwise known as 6 plus 5.

[8] Interview with author, as are all direct quotes in this chapter from Jack Copeland unless stated.

Pilot ACE undergoing testing, circa 1949
Image: courtesy of the Computer History Museum, CC BY-NC-SA

with that instruction table, from then on, you never need to change the setup of the scanner again. You just write instructions on the tape and the machine will follow them."

It's a concept that presages stored program computers almost 15 years before they came into existence. Specifically, Turing's universal machine acts as the theoretical blueprint for the ACE that he would design a decade later: a generic machine that can be programmed using coded instructions stored in memory.[9] With the ACE, the tape's place is taken by mercury-delay lines.

Through Newman's influence, by the time Turing's paper was published he had started two years' placement at Princeton University, where he would rub shoulders with the brilliant John von Neumann. The latter would one day be called the father of computing for his work on the EDVAC, but even by this point in history he was

---

[9] Turing's 36-page paper is available to download from **cs.virginia.edu/~robins/Turing_Paper_1936.pdf** and other sources if you wish to follow the intricate mathematics involved.

considered a groundbreaking theorist thanks to a string of influential papers covering everything from game theory to quantum mechanics.

During his two years at Princeton, Turing completed his PhD thesis with support from Alonzo Church and worked with von Neumann on group theory. We also know that von Neumann was aware of Turing's Computable Numbers paper because he cited a key concept from it in a lecture series[10] (probably delivered in 1945).

Some have supplied more detail. "Von Neumann introduced me to ['On Computable Numbers'] and at his urging I studied it with care," wrote von Neumann's Manhattan Project colleague Stan Frankel in 1972.[11] "Many people have acclaimed von Neumann as the 'father of the computer' (in a modern sense of the term) but I am sure that he would never have made that mistake himself. He might well be called the midwife, perhaps, but he firmly emphasised to me, and to others I am sure, that the fundamental conception is owing to Turing – insofar as not anticipated by Babbage, Lovelace, and others."

Whilst at Princeton, Turing moved from theory to practicality by designing an electric multiplier. A physics graduate called Malcolm MacPhail then lent Turing a key to one of Princeton's labs and taught him "how to use the lathe, drill, press etc. without chopping off his fingers," wrote MacPhail to Turing's biographer, Andrew Hodges.[12] "And so, he machined and wound the relays; and, to our surprise and delight, the calculator worked."

According to MacPhail, the two "had many discussions" on the subject of cryptography in the year leading up to his departure. Further evidence of Turing's interest in the subject comes from letters he sent at that time. "Even at Princeton, [Turing] wrote to his mother and said he had devised a code that he thought was pretty well impossible to break," says Copeland. "And he kind of joked about selling it to the British government."

---

[10] Thomas Haigh and Mark Priestley, 'Von Neumann Thought Turing's Universal Machine was "Simple and Neat"', Communications of the ACM, 1 January 2020, **rpimag.co/simpleandneat**
[11] Brian Randell, 'On Alan Turing and the Origins of digital computers', part of a *Technical Report Series* (number 33) edited by Dr B Shaw, University of Newcastle, May 1972, p15. Randell wrote to Dr Frankel to find out more about von Neumann's awareness of Turing's paper, and the quote is directly lifted from Randell's transcript of his reply. A copy of Randell's report is held at the British Library, London.
[12] Andrew Hodges, *Alan Turing: The Enigma* (Vintage Books, 2014, paperback edition, ISBN 978-1784700089), p175

Turing rapidly discovered that the government had their own plans for mathematicians such as him. After gaining his PhD in June 1938 and returning to Cambridge, he attended a cryptology course run by the UK's Government Code and Cypher School (GC&CS). Copeland's research show that "Turing started regularly visiting the London offices of GC&CS at the beginning of 1939, and he started talking to Dilly Knox[13] about Enigma."

Throughout this time, Turing remained a fellow at the University of Cambridge. He became interested in the Riemann zeta function, another legendary unsolved problem (this time dating back to 1859). Turing devised a new and theoretical approach to calculating the function, but to complete his work he needed to perform calculations well beyond the desktop calculators of the day. Or the university's model differential analyser, at that time under the control of Maurice Wilkes, creator of EDSAC. So, naturally, Turing set to work designing his own machine.

He estimated this would cost £50, almost £3000 in today's money and well beyond his means, so he applied for and was given a grant by the Royal Society. Together with the enthusiastic help of Donald MacPhail, brother of Malcolm, who happened to be a research student at Cambridge studying mechanical engineering, he set to work on the schematics. By the summer of 1939, a visitor to Turing's rooms would have been met by a collection of gear wheels. A completed machine would need 80 of them.

But war intervened and the project was destined to never be completed. Instead, Turing and his brilliant colleagues at Bletchley Park set to work on an entirely different machine. One that could help decipher the Enigma-encrypted radio transmissions of the German military.

The British code breakers had been given an invaluable head start by the work of Polish cryptanalysts, who had created a machine they called 'bomba'. While it's natural to assume that it was named after bombs – 'bomba' is Polish for bombs (although it also translates as 'impressive') and the mechanisms made a muffled, rhythmic sound – Copeland provides a more interesting explanation. "Marian Rejewski, who with Anthony Palluth was the chief architect of the bomba, has been quoted as saying that when they needed a code name for these things, he happened to be eating a bomba

---

[13] Alfred Dillwyn Knox famously helped to decrypt the Zimmermann Telegram, pivotal in making the USA join World War I, and was a central figure at Bletchley Park until his death, aged 58, from cancer in February 1943.

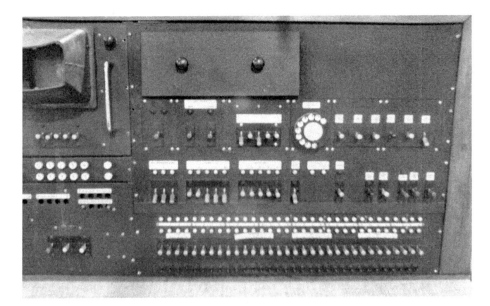

Pilot ACE console
Image: Karl Baron, CC BY 2.0

at the time. He was sitting in a café eating an ice cream, and he said let's call the machine a bomba."

The Polish team had created six bomby (pronounced 'bom-bee') by the end of 1938. There were six to match the number of possible wheel orders of the Enigma, and each bomba cycled through 17,576 permutations to discover the right setting. However, the machines' usefulness was reduced by a factor of ten when the Germans increased the number of wheel orders to 60.

"It wasn't just the extra wheels that made life tougher," says Copeland. "What really brought them down was the 'Stecker', the plugboard, because the Germans increased the number of steckered letters." Essentially, the Stecker board scrambled the letters, making it even harder to decrypt messages. While the first version of the Stecker board only worked on eight letters, this was increased to a dozen, which was enough to defeat the bomby.

All of this meant that the Poles could no longer work on their own. At the same time, the chances of Nazi occupation grew – Germany had invaded Czechoslovakia on 14 March 1939 and Hitler's speeches increasingly targeted Poland – which forced the Polish code breakers to reveal their hand. They had to share all they knew with their allies, namely France and Britain. So, in late July 1939, Dilly Knox along with two other British representatives found himself face to face with the Poles at their Warsaw headquarters. It was here he learned about the existence of the six bomby.

Knox left Warsaw deeply impressed by the Poles' work – even sending them ribbons for "winning" the race to decrypt the German Enigma – but knew that the British had a huge task on their hands. They needed to create something an order of magnitude more powerful than the bomba. Mavis Batey, one of the code breakers who worked with Dilly Knox and author of *Dilly: The Man Who Broke Enigmas*, quotes his post-meeting report:[14] "At the time of my visit I had ideas which seemed to be better, and I have since discussed them exhaustively with Mr Turing and Commander Travis," wrote Knox. He believed, the report states, that they could produce a machine that they wanted to call the "Geschlechtzylinder". A name, luckily, that never caught on.

While Knox was a gifted code breaker, he recognised that he needed help from a mathematician and that Alan Turing was the right man for the job. His ideas on logic and computability had been honed over the past five years with the likes of Newman; he already had an affinity for cryptoanalysis; and he had hands-on experience with relays and machine design.

In her excellent chapter 'Breaking machines with a pencil', found in *The Turing Guide*,[15] Batey describes how Knox gave Turing full responsibility for developing the machine as summer turned to autumn. "[Turing] worked in the stable-yard cottage that Dilly had chosen to be away from the administrators," she wrote. To be precise, the cottage's loft. "Since the only access was a ladder in the wall, two of the girls rigged up a pulley and a basket for sending up coffee and sandwiches," Batey added.

Thus fuelled, an inspired Turing set to work. By 1 November 1939,[16] he was able to set down the requirements in detail. The job of building the British bombe then passed to Harold 'Doc' Keen of the British Tabulating Machine Company, and within six months he delivered: the first British bombe, 'Victory', was installed and ready for action by the end of April.

---

[14] Mavis Batey, *Dilly: The Man Who Broke Enigmas*, (Biteback Publishing, Kindle Edition), p119
[15] Various authors, *The Turing Guide* (Oxford University Press, 2017, ISBN 978-0198747826), pp97-107
[16] As above, p105

It was a huge beast, standing six feet and six inches tall, over seven feet long and three feet wide. But it wasn't merely the size that made it impressive, as this description from Turing's Bletchley Park colleague Patrick Mahon makes clear: "From one side, a bombe appears to consist of nine rows of revolving drums; from the other, of coils of coloured, reminiscent of a Fair Isle sweater."[17]

The bombe was an electromechanical machine designed to simulate the Enigma's wheels and circuits. It can be most easily thought of as multiple simulated Enigma machines, all hooked together.[18]

The system relied on 'cribs', expected words or phrases such as 'Wetter fuer die Nacht' ('weather for the night' in English). We won't go into detail here, but in essence the bombe would search the space of possible settings of the Enigma and, when a group of settings produced an output that matched an expected pattern, it would stop. The operators would then note down the settings, which were then transferred to a replica Enigma machine. If this produced German, albeit peppered with wrong letters, the bombe had cracked the code. Time to move on to the next message.

Unfortunately, Turing's bombe only worked if the crib met certain criteria; in particular, it needed to include three 'closed chains' or loops.[19] It took the insight of another talented mathematician, Gordon Welchman, to turn what would have been an effective machine into a juggernaut. Welchman's insight was to spot that due to the Stecker board's reciprocal nature – that, if A turned into B, B would turn into A – they could simultaneously scan all the possible Stecker values. Welchman designed a 'diagonal' board to take advantage of this, and by August the first of the new, improved bombes came into service.[20]

From this point on, the British Tabulating Machine Company ramped up production, eventually building a dedicated bombe factory. "We were taken there to watch them being made and to encourage the workers, although we thought their

---

[17] Patrick Mahon, 'History of Hut 8 to December 1941', in Jack Copeland (ed.), *The Essential Turing*
[18] This is a much-simplified version of how the bombe worked. For more detail, read *The Essential Turing*, pp246-254.
[19] 'Closed chains' is Turing's word, but Jack Copeland uses the friendlier term of 'loops' in *The Essential Turing*, where he explains the process in detail (see pp120-123). Not all cribs produced the required three loops, which is why Turing's first bombe design was less effective.
[20] By necessity, we have missed out much detail here. Interested readers have numerous books to choose from to discover more about the British bombes, including Andrew Hodges' biography, *Alan Turing: The Enigma*, *Codebreakers: The Inside Story of Bletchley Park* edited by Alan Stripp and Sir Francis Hinsley, and *The Essential Turing* edited by Jack Copeland.

conditions were better than ours," wrote bombe operator Diana Payne.[21] "It was a surprise to see the large number of machines in production."

At first the bombes were housed at Bletchley Park, but soon they became too numerous and new machines were shipped to other secret locations (including Woburn Sands, Eastcote, and Stanmore). In an early example of remote computing, Bletchley Park operators could phone in commands, and eventually they could even call up US-made bombes in Washington, DC.

The bombes needed hands-on control too. Much of this work was done by members of the Auxiliary Territorial Service, or ATS, the women's branch of the British Army founded in 1938. Or by the Women's Royal Naval Service, shortened to the Wrens. "We were given a menu, which was a complicated drawing of numbers and letters from which we plugged up the back of the machine and set the drums on the front," wrote Payne.[22]

"We only knew the subject of the key and never the contents of the messages," she added. "All this work had to be done at top speed, and at the same time 100 per cent accuracy was essential. The bombes made a considerable noise as the drums revolved, each row at a different speed, so there was not much talking during the eight-hour spell. For technical reasons which I never understood, the bombe would suddenly stop, and we took a reading from the drums."

But Turing's wartime work was far from done. He was in charge of Hut 8, which focused on Naval Enigma, and these messages were the toughest to break due to an extra layer of encryption. "They doubly enciphered the indicators," explains Copeland. "They would encipher them once by hand, using a book of settings, and then they enciphered them again using the Enigma machine."

There was one particular Enigma network of concern: Dolphin. This was the code name given to U-boats' communications, and with the fall of France in June 1940 came the menace of growing U-boat attacks in the north Atlantic. And they were deadly, with one U-boat alone – U-48 – responsible for sinking 51 ships.[23] Almost 600 merchant ships had been sunk by the end of 1940.[24]

---

[21] Diana Payne, 'The bombes', Chapter 17 of *Codebreakers: The Inside Story of Bletchley Park*, edited by Alan Stripp and Sir Francis Hinsley (Thistle Publishing, Kindle Edition), p169
[22] As above, pp169-170
[23] According to **uboat.net, uboat.net/boats/u48.htm**
[24] Jack Copeland (ed.), *The Essential Turing*, p257

Even with bombes to call upon, Hut 8 required help if it was to break Dolphin. The code breakers needed information direct from the German ships and U-boats using Dolphin, and that meant audacious raids – or 'pinches' as they were called. Those with a taste for adventure should read *Enigma: The Battle for the Code* by Hugh Sebag-Montefiore,[25] which details many of the most daring achievements.

There were several partially successful pinches, where the Allied forces would grab whatever information they could: records of previous messages that could act as cribs, daily codes, Stecker settings.

But the real prize was the current bigram tables, used when doubly enciphering the indicators. Every Enigma operator was issued a set of nine bigram tables that told them what to do with each pair of letters (each bigram): for example, that 'AC' should become 'QS'. In total, the tables included 676 bigrams. The same tables were used for months at a time, which was another reason they were so valuable.

The problem was that the Germans also knew how important this documentation was, and naval commanders were under strict orders to destroy it all as soon as they came under attack. So, while several pinches happened during 1940, pickings were slim. As summer turned into autumn and the U-boats continued to sink Allied ships, the Admiralty became open to the most outlandish of ideas. Including one from the man who would later create James Bond.

Commander Ian Fleming devised 'Project Ruthless', and it lived up to its name. The key idea was for a daring crew of five, dressed in Luftwaffe uniform, to crash a German bomber (obtained from the Air Ministry) in the English Channel. When rescued by the Germans, they would "shoot German crew, dump overboard, bring rescue boat to English port".[26] Not the most sporting of approaches, and it was never put into practice.

"Turing and Twinn came to me like undertakers cheated of a nice corpse two days ago, all in a stew about the cancellation of Operation Ruthless," wrote Frank Birch, who was in charge of Bletchley's naval interpretation section, in a memo to the Admiralty.[27] "The burden of their song was the importance of a pinch. Did the authorities realise that … there was very little hope, if any, of their deciphering

---

[25] Hugh Sebag-Montefiore, *Enigma: The Battle for the Code* (Weidenfeld & Nicolson, 2000, ISBN 978-1474608329)
[26] 'Operation Ruthless, October 1940', **turing.org.uk/sources/ruthless.html**
[27] As above. Peter Twinn worked with Turing in Hut 8.

A Bletchley Park bombe, circa 1945
Image: Public Domain

current, or even approximately current, Enigma for months and months and months – if ever?"

A long-awaited breakthrough came in March 1941, when a Royal Navy raid on a German trawler off the Norwegian coast yielded the previous month's daily keys. This gave Turing the information he needed to build the bigram tables mentioned above, and now, at last, they could break into Dolphin. Further successful raids meant they could repeat the feat when the Germans issued new bigram tables later that year.

But the course of war never runs smooth, as Hut 8 discovered on 1 February 1942 when the four-wheel version of the Enigma used by the Kriegsmarine went into general use. This meant the bombes had to cope with 26 times more options, and there was simply no way to create 26 times as many bombes. Work began on a four-wheel version of the bombe, among other approaches, but these efforts were more mechanical than theoretical so Turing was no longer central to the work.

The four-wheel problem was instead solved – or at least salved, as the Germans continued to add new security methods for the rest of the war – by the capture of a U-boat, and the code books it contained, on 30 October 1942. Only a few days later, Turing would sail to the United States to better co-ordinate the two countries' code breaking efforts and help the US build their own bombes at a factory in Dayton, Ohio.

His other task was to work with the Americans on their speech-encipherment system, dubbed SIGSALY. This was a room-filling beast being developed at Bell Labs, where George Stibitz built his relay computers, subject of Chapter 3. Turing's time at Bell Labs brought him into contact with Claude Shannon, mainly through lunchtime chats. Shannon wasn't involved with the speech encryption, but he shared Turing's interest in creating computing machines and would go on to found what we now call information theory in a 1948 paper called 'A Mathematical Theory of Communication'.

"Turing and I had an awful lot in common," said Shannon in 1982.[28] "And we spent much time discussing the concepts of what's in the human brain... how the brain is built, how it works, and what can be done with machines and whether you can do anything with machines that you can do with the human brain and so on."

When Turing returned to Bletchley Park in March 1943, breaking naval Enigma messages had become almost routine again. Construction of Colossus was under way at Dollis Hill. It was time for a new challenge.

So Turing turned his attention to the problem of miniaturising a speech encryption machine. He spent the next year devising the system and then, with the help of electrical engineer Don Bayley, manufacturing the machine itself. A device they called Delilah for its deceptive ways.

By the time VE Day finally arrived, on 8 May 1945, Turing and Bayley had produced a working prototype of Delilah. Unlike the American system, Delilah was a portable unit, but it arrived too late and was never put into active service.[29] Nevertheless, it's another example of Turing's inventive genius and suggests that

---

[28] 'Claude E Shannon, an oral history' conducted in 1982 by Robert Price, IEEE History Center, Piscataway, NJ, USA. **rpimag.co/shannoninterview**

[29] A replica of Delilah is available to view at Bletchley Park.

he was in the right frame of mind to tackle an even bigger project when the opportunity arose.

That ambition was essentially to build an electronic brain. In the universal machine described by his 1936 paper, he had arguably created a logical structure equivalent to everything that takes place in our physical brains.

Then, a month later, came an invitation out of the blue. Through Newman, John Womersley met Turing and offered him a job: to build a computer. This, according to Copeland, was the chance Turing had been waiting for: "Tom Flowers told me that once Turing, and also Newman, saw Colossus, it was just a question of their waiting for an opportunity to build a universal Turing machine in hardware, and along it came in the form of Womersley. So Turing just leapt on it."

Womersley was newly installed at the National Physical Laboratory (NPL) in Teddington, west London, which was pivotal in both inventing radar and developing systems to detect aircraft (although much credit is due to Sir Robert Watson-Watt alone). The NPL also later earned fame for creating the 'bouncing bomb' used in the Dambusters raid. Now that Bletchley Park was disbanding, it made sense for the NPL to be the country's new centre of excellence for computation.

Womersley's new territory at NPL, called the NPL's Mathematics Division, had five pillars. The General Computing Section would tackle computational problems, with support from the Punched Card and Differential Analyser Sections, with a separate section dedicated to statistics. Turing took on the fifth section, dedicated to creating a new computer.

After a few weeks spent in Germany examining what its scientists had achieved in the war, Turing took up his post.[30] By this time, he had read von Neumann's draft report on the EDVAC, but unlike Maurice Wilkes – creator of the EDSAC – he did not use the EDVAC as the basis of his design. He wanted to turn his universal machine into a buzzing electronic reality based on his own ideas.

What happened next was Turing at his absolute best. Left alone, with no distractions, he wrote a detailed report simply titled 'Proposed electronic calculator' in the space of three months.[31] The first section of the report covered the device's

---

[30] Tommy Flowers accompanied Turing on this trip, which took place from mid-July to mid-August. Sadly, they did not encounter Konrad Zuse (see Chapter 2) during the investigation.

[31] You can read the full report on AlanTuring.net at **alanturing.net/turing_archive/archive/p/p01/p01.php**

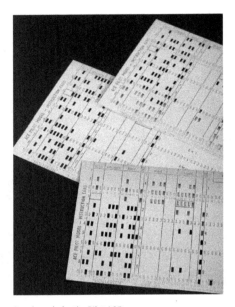

Punch cards for the Pilot ACE
Image: Science Museum London, CC BY-SA 2.0

composition, including the type and amount of storage required, descriptions of the logical circuits, and example problems it could solve. The second section went into greater detail, including several pages dedicated to the theory of delay line storage. An appendix set out detailed flow diagrams to show how the logical operations worked. It was effectively a blueprint of how to build a universal computing machine.

But it was Womersley who gave the computer its name: the Automatic Computing Engine. This was a nod to Charles Babbage's Analytical Engine, but also translated into a friendly acronym: ACE. The two men combined effectively in a crucial meeting to the Executive Committee of the NPL in March 1946.[32]

Womersley's opening pitch was that the ACE had the potential to be greater than the sum of the ENIAC, Harvard Mark 1, and Bell Labs computers. To be superior in both output and versatility. And that in Alan Turing, they had the man for the job. Turing then took over, providing a detailed breakdown of the proposed computer – both its applications and the challenges that stood in their way.

The minutes also make clear that Professor Douglas Hartree, who probably knew of the still-secret Colossus, was an enthusiastic backer of the project. He pointed out that Turing's design was far more efficient than the ENIAC's, needing around 2000 valves rather than 18,000 and yet having a vastly superior storage capacity: 6000 numbers versus 20. "Professor Hartree also pointed out that if the ACE is not developed in this country, the USA will sweep the field," the minutes close.

---

[32] The minutes of the meeting are published in *Alan Turing's Automatic Computing Engine*, edited by Jack Copeland (Oxford University Press, 2005, ISBN 978-0198565932), pp 47-53

The discussion soon moved on to money. Rather than build the full machine, Womersley proposed that a pilot could be created for "approximately £10,000". This was a small enough sum that it could be found from the NPL's existing budget, although the full machine might cost ten times as much. This "should be regarded as a contribution of the Laboratory to the general good of the country" read the minutes, attributing the comment to Hartree once more.

We can imagine the patriotic stirrings among the attending men at this moment, made all the more vivid by a statement that the "Committee resolved unanimously to support with enthusiasm the proposal".[33] Over the coming three years, that enthusiasm was to be sorely tested.

This was despite good early progress. Turing continued work on the ACE's design while the NPL delegated the engineering work to a Post Office team led by Tommy Flowers. It was an approach that had worked exceptionally well with the Colossus, after all. But the Colossus was built during wartime, with a sense of existential urgency; in the post-war period, the Dollis Hill team's priority was to tackle a backlog of Post Office work stemming from the war. It couldn't give the ACE project anywhere near the same attention as it had Colossus.

Perhaps this should have been obvious to the NPL, but a report dated 7 March 1946 from the Post Office suggested that the team would have a "minimal ACE by August or September".[34] This despite the fact that Flowers could only spare two men for the project, William Chandler and Allen Coombs, and as summer turned into autumn with no sign of hardware it finally became clear to the NPL management team that they needed to think about alternatives.

Hartree, who was also a professor at the University of Cambridge, suggested that Maurice Wilkes might be able to help. By this time, he and his team were making steady progress on EDSAC, including the manufacture of delay lines, and Hartree reported to the committee that Wilkes was "prepared to give as much help as he could". But while Wilkes was following von Neumann's EDVAC recipe to make a computer, Turing had entirely different ideas. How different is evident from Turing's

---

[33] There was even talk of building duplicates, which Womersley estimated would cost around a quarter of the first ACE. But he then poured cold water on the idea, suggesting that "the capacity of the machine would be such that duplicates would not be needed". In other words, one such powerful computer would be enough for the whole country.

[34] An extract of the memo is published in *Alan Turing's Automatic Computing Engine*, edited by Jack Copeland, p50

description of one aspect of Wilkes' proposals as "rank folly".[35] Nothing came of the proposed partnership.

While engineering progress was effectively zero in 1946, the mathematical side raced ahead. In May, James Wilkinson joined forces with Turing – who had worked alone until then – to create a formidable intellectual pairing. While Wilkinson had no prior knowledge of computers, he had graduated from Cambridge as Senior Wrangler. That is, the highest scorer of the entire year's mathematics graduates. Later that summer, the pair were joined by another maths graduate, Mike Woodger.

The two newcomers, once brought up to speed on what this new-fangled electronic computer could do, spent most of their time developing 'instruction tables' for the non-existent hardware. By the end of 1946, the team had created instruction tables (programs or even subroutines in more modern parlance) for division, square root, sine and cosine, logarithms, sets of linear equations and many more.[36]

During 1947, Gerry Alway and Donald Davies joined the group, plus the sole woman on the team, Betty Curtis. "Woodger described her as the Grace Hopper of British computing," says Copeland. He adds: "One reason why the Pilot ACE was so terrific straight away was because these chaps and Betty Curtis were sitting there writing the programs, while the hardware people were failing to build anything."

Turing, meanwhile, concentrated on the ACE's logical design. By the time Wilkinson and Woodger arrived in the summer of 1946, he had developed his initial design to what appears to be a fifth iteration – Turing dubbed it version V. But Turing was always on the hunt for ways to increase the speed of operation, so by the end of the year he was on version VII, which Wilkinson described as a "very significant advance"[37] in terms of logic over version V.

This moving of the goalposts, as anyone who has worked on similar projects will know, is challenging. It appears to be another reason why the partnership with the Post Office didn't work, with Coombs – one of the two engineers working under Flowers – saying "every time we went [to the NPL] and said 'Right, now! What do you want us to make?', we'd find that the last idea, that they gave us last week, was

---

[35] Jack Copeland (ed.), *Alan Turing's Automatic Computing Engine*, p62
[36] NPL Report for the Year, 1946, **alanturing.net/turing_archive/archive/l/l76/L76-011b.html**
[37] 'The Pilot Ace (1982)', interview featuring Mike Woodger and Jim Wilkinson, located on NPL's YouTube channel: **youtu.be/Sf28IJmm-P4**

old hat and they'd got a quite different one".[38] As the most extreme example, at one point the Post Office engineers were told that the NPL was considering switching to the cathode ray tube storage idea being developed in Manchester, which would have made all their work on mercury storage delay lines pointless.

Ultimately, and almost certainly to the relief of both partners, the NPL decided to part ways with the General Post Office. This was on the understanding that the GPO could continue development of its own version of the computer separately. Which, as we will discover at the end of this chapter, it did.

Such was the unsatisfactory state of affairs when Harry Huskey joined the NPL on a one-year visit on 4 January 1947. This was arranged by Hartree, who had met Huskey during his tour of American computers in 1946. Unlike any of the others on the Pilot ACE team, Huskey had hands-on experience with computer hardware, having joined the ENIAC project in 1944 and eventually writing the computer's operation and maintenance manuals.

Huskey wasn't willing to sit around writing routines for a non-existent computer. In April, he persuaded Womersley that they could build a 'Test Assembly' version of the hardware themselves rather than wait for outsiders. This would be the first time that they had taken an integrated approach, with one team creating the logical design – a stripped-down version of Turing's Version V – and the physical apparatus.

By now, Turing shared the belief that the theory and physical sides of the project should happen in the same place. A belief cemented by his visit to the United States in January 1947 to see how their various computing projects were progressing. "One point concerning the form of organisation struck me very strongly," he wrote in his report.[39] "The engineering development work was in every case being done in the same building with the more mathematical work. I am convinced this is the right approach."

Even on his return from America, Turing spent time away from the NPL, with Woodger saying that he was often at King's College in Cambridge or visiting the University of Manchester.[40] "[When] he came back from one of these occasions, he

---

[38] Jack Copeland (ed.), *Alan Turing's Automatic Computing Engine*, p38
[39] Alan Turing, 'Report on visit to U.S.A., January 1st-20th, 1947',
alanturing.net/turing_archive/archive/p/p20/p20.php
[40] 'The Pilot Ace (1982)', interview featuring Mike Woodger and Jim Wilkinson, located on NPL's YouTube channel:
youtu.be/Sf28IJmm-P4

didn't know of this work that Huskey was doing, and he looked over my shoulder to see what I was doing," said Woodger. "I had written a preliminary piece of program for this new Huskey machine. And of course, I gave it a name, version H, and he looked over my shoulder and he said, 'What's this version H?' Then it transpired that behind his back, Huskey had the nerve to try to design a computer."

Turing's immediate reaction was predictably furious: quite aside from this happening without his agreement, he likely felt it was a poor use of his team's time. While he never played an active role in the Test Assembly, he did lend his support when requested. For example, in a 2006 interview, Huskey remembered asking Turing about the possibility of reflections from sound waves in the delay lines, thinking they would have to build a line to test this. Turing instead "got out his pencil and paper and worked out the mathematics and said that he didn't think there was a problem".[41] This wasn't unusual behaviour, Huskey added: "[If] somebody had a little problem and they were having a terrible time, he would … go way out of his way to help them."

Huskey, Wilkinson, and Woodger made slow but steady progress on the Test Assembly, completing circuit designs, commissioning a mercury delay line from the NPL machine shop, and even hunting out valves that met their specification. "Wilkinson and I went around to military dumps where at the end of the war the various military operations had all kinds of parts left over, from bomb sights to you name it," said Huskey.

By October, there was a whispered hope that the Test Assembly would be complete before Huskey departed at the end of the year – although Huskey believed it would have more likely been at some point in 1948 – only for the head of the NPL, Sir Charles Darwin (grandson of his more famous namesake), to halt the Test Assembly project.

The seeds for this decision had been planted at the start of the year, in light of Turing's suggestion that engineering should happen in-house at the NPL as it did in America. The most logical home for this project was within the NPL's Radio Division, Darwin felt, and the opportunity to lead the team was eagerly seized by one of that division's managers, Horace Augustus Thomas. But there were soon signs

---

[41] 'Oral History of Harry Huskey' by William Aspray. Recorded 7 February, 2006, Computer History Museum. See computerhistory.org/collections/catalog/102657983, p22

that Thomas was "behaving as if he is starting up a new Division".[42] Not least when Thomas's team removed the bulk of the Test Assembly's equipment while Huskey and Turing both happened to be on holiday.

It was Thomas who persuaded Darwin to close the Test Assembly project down, leaving the project in disarray. Turing's team had already been reduced, with two members transferred to the Radio Division. Worse still, Thomas was determined to start afresh – despite his team's lack of knowledge about logic circuits – rather than work with Huskey, Wilkinson, and Woodger.

"Darwin, for all his virtues, just didn't know what he'd got," says Copeland. "If Thomas had not existed, and Darwin had been a different kind of administrator, I think Version H might well have flourished." In particular, it could have been the world's first stored-program computer, a title instead claimed by the Manchester Baby (see Chapter 7). "That was the barest of electronics," says Copeland. "One Williams tube and a few capacitors and they were away – and they've gone down in history."

The most dispirited team member of all turned out to be Alan Turing, who took sabbatical leave in late 1947 so that he could dedicate himself to machine intelligence research back at Cambridge. In July that year, Darwin rather optimistically wrote that the ACE "has now got to the stage of ironmongery"[43] and evidently was happy for its inventor to be absent for an extended period.

Turing wasn't the only one looking for new challenges. In February 1948, Thomas surprised everyone by accepting a job at Unilever, leaving the Pilot ACE without any obvious leader. At this point, the whole project might have fallen away, a giant bullet point heading the list of 'IT projects that failed'. Fortunately, providence came in the form of Francis Morley Colebrook.

A man in his mid-50s, Colebrook took a mature, collegiate approach, which may be one of the key reasons he was chosen to replace Thomas in March 1948 as head of the newly formed Electronics Section of Radio Division. He soon surprised Wilkinson by suggesting that the two warring groups should join forces and physically work together in the same lab. Wilkinson, once he'd recovered from his shock, agreed.

---

[42] Letter from ES Hiscocks, secretary of the NPL, to Charles Darwin, 12 August 1947, reprinted in *Alan Turing's Electronic Brain* by Jack Copeland and others (Oxford University Press, 2005, ISBN 978-0199609154), p68
[43] Letter from Sir Charles Darwin to Sir Edward Appleton, 23 July 1947,
alanturing.net/turing_archive/archive/p/p30/p30.php

From this point on, the project progressed as smoothly as could be hoped. Colebrook's team benefited from the newly imported mathematical brains of Wilkinson, Woodger, Alway, Curtis, and Davies; the mathematicians learned better electronics techniques from experts such as Ted Newman and David Clayden.[44] Finally, the stars were starting to align.

There were numerous problems to solve, with Newman and Clayden dedicating much time transforming Turing's logic circuit designs into something that would work efficiently based on the latest techniques. Meanwhile, building reliable mercury delay lines continued to be a challenge.

It wasn't until the start of 1949, wrote Wilkinson, that they "started on the detailed design of the Pilot ACE".[45] It took most of the remainder of the year for the team to finish the plans and for the NPL workshops to build them. Assembly began in the autumn, and slowly the Pilot ACE took shape. "It was finally in May of 1950[46] that we had sufficient of it assembled for the machine to work, in the sense that it stored the program in its memory and it executed that program," said Wilkinson.

By this time Darwin had left, replaced by Edward 'Teddy' Bullard, who earned his scientific reputation for developing and refining the idea of a geomagnetic dynamo controlling the Earth's magnetic fields.[47] Bullard visited Wilkinson soon after becoming director and was sceptical when told that the Pilot ACE would be working any day.

"He didn't really believe me," said Wilkinson.[48] "And I persisted and he said, 'Well, as soon as it works, you must call me.' And so when it did work, I tried to get in touch with him, and of course, we couldn't get him. And so I paced up and down, saying, 'Where's the bloody director? You can never find him when you want him.' And I said this several times, and just as I was about to say it again, he stepped in through the window and said, 'Well, here's the bloody director.'"

---

[44] Newman and Clayden had recently joined the NPL from Electric and Musical Industries Limited, better known as EMI. By the late 1940s, this had grown from its gramophone roots to become a manufacturer of radar and broadcast equipment. While Horace Thomas is painted as a villain in certain accounts, he deserves credit for hiring both Newman and Clayden. "I thought he'd made a very impressive place to work in," wrote Newman of Thomas in issue 9 of *Resurrection, The Bulletin of the Computer Conservation Society* (ISSN 0958-7403, spring 1994). "I say that because a lot of people have derided him a great deal, and I don't think it was fair."

[45] Jack Copeland (ed.), *Alan Turing's Automatic Computing Engine*, Chapter 4, p98

[46] To be precise, Wednesday 10 May 1950.

[47] Equal credit should go to Walter Elasser.

[48] 'The Pilot Ace (1982)', interview featuring Mike Woodger and Jim Wilkinson, located on NPL's YouTube channel: **youtu.be/Sf28IJmm-P4**

Wilkinson demonstrated the program, but at this point the Pilot ACE was "in a very primitive state" so all he could show was that it could successfully execute basic arithmetic. "He was obviously very pleased and rather surprised that it was working," said Wilkinson, "but he turned to me, and he said, 'The program's less than epoch-making.' And that, I had to admit, was true."

From this point onwards, Wilkinson and the rest of the team were under pressure to develop the Pilot ACE to the point where they could demonstrate it to the press and their industry peers. They would need something more impressive than primary school mathematics. This process took another six months, until finally, on 29 November 1950, the computer was ready to make its first public appearance – a little over five years after Turing had started work on the project.

Wilkinson describes the "ordeal" of a three-day demonstration to the national press, technical press and then VIPs, with good humour. "To be perfectly frank, the machine had never worked really well up to that time, and the chances that it would give a flawless or even an acceptable demonstration over those three days was really rather small. And to everybody's amazement, and more to ours than anybody else, because we knew its rather shaky condition, it proceeded to give the most impeccable performance."

It certainly impressed the BBC, which broadcast a two-minute newsreel entitled 'THE ACE CALCULATOR' on 1 December 1950.[49] "Its details are truly astonishing," says the well-spoken narrator. "It consists of 800 valves. Its running costs are about a pound a minute, but in that time, it can work out more sums than a hand-cranked machine could do in a month. For example, to find whether 99,997 is a prime number … one takes just a set of cards, feeds them into the machine, and within a few seconds, the answer comes up."

The Times had published a similarly glowing report the previous day.[50] "The Ace itself will be built later," states the article, "but the model demonstrated here to-day is nonetheless a complete electronic calculating machine, claimed as one of the fastest and most powerful computing devices in the world." It adds that this speed "could

---

[49] 'THE ACE CALCULATOR', BBC Archive (Facebook), first broadcast 1 December 1950 and republished 1 December 2019, **facebook.com/share/v/18x8Mk6HN2**
[50] 'Month's work in a minute', *The Times*, 30 November 1950

perhaps best be grasped from the fact that it could provide the correct answer in one minute to a problem that would occupy a mathematician for a month".

The article ends optimistically but wrongly: "Now it [the Pilot ACE] is ready to 'do business' and is expected to more than earn its keep." In fact, it would take the whole of 1951 – including a redesign of the delay lines and the addition of a multiplier unit – before the machine was ready to be dismantled and rebuilt in its proper home within the NPL's Mathematics division. This happened in February 1952, at which point the Pilot ACE was indeed put into service.

Profitable service too. It essentially became a computer for hire, there to tackle complex problems that the NPL's differential analyser and fleets of desk calculators couldn't easily handle. Famously, it helped to solve the mystery of why a de Havilland Comet plane crashed in January 1954, killing 29 passengers and six crew. Rather than sabotage, as initially suspected, the Pilot ACE team took raw information from the investigating team and then used this to determine what had gone wrong. "Eight million multiplications was a small part of the job," said Tom Vickers, one of the computer's first operators.[51] "And eventually we found the answer to the Comet disaster."

By combing through official records,[52] Copeland estimated that the Pilot ACE earned the NPL around £100,000 over its working life – £24,000 in 1954 alone – which was an excellent return on a machine that cost less than £50,000 to build.

"This was big money as far as the lab was concerned," said Vickers. "Once ... people got to hear about it, and this was the only machine around, we just filled up." And clearly the Pilot ACE was worth the high cost, because a multiplication that might have taken 15 seconds on a desk calculator could now be done in a 500th of a second.

It's worth emphasising just how fast the Pilot ACE was. Turing put speed of operations at the heart of his design, using coding to minimise wait times. This became known as 'optimum coding', although Turing never used the term.

"The instructions needed to pop out of the delay line at exactly the right moment," explains Copeland. Take a program that, at some point, requires values A and B to be

---

[51] 'Piloting Computing – Alan Turing's Computing Engine', documentary located on NPL's YouTube channel: youtu.be/cEQ6cnwaY_s
[52] Jack Copeland (Ed.), *Alan Turing's Automatic Computing Machine*, p74

added together. "So A pops out of its delay at a particular time, and B pops out of its delay line, hopefully at the same time. With optimum coding, the instruction to add them needs to pop out of its delay line synchronously with A and B."

For that to happen, the programmer must place the instructions in the right places in the delay lines, explains Copeland. "Otherwise you'd need some 'wait' commands, where if the instruction came out too early then the machine would need to waste a few cycles of time before obeying the instruction."

While optimum coding was incredibly clever, and in many ways ahead of its time, it also added an extra layer of complication to programming. "It was a total nightmare," says Copeland. "I'm surprised they kept sane while they were doing it."

The finished machine ran at 1MHz, twice the speed of the EDSAC, and included two arithmetic units: one supported 32-bit numbers, accurate to roughly nine decimal places, the other for longer 64-bit numbers. While Turing's initial design for the ACE called for 200 mercury delay lines, the Pilot version included a rather more modest dozen. This limited storage to 768 words, but the team got around this by adding four slower but far more capacious magnetic drums in 1954, bringing an extra 32,768 words into the Pilot ACE's memory. The downside: it would take up to 7 milliseconds (ms) to read them, roughly 20-30 times slower than a mercury delay line (which refreshed every 200 to 300 microseconds, or 0.2 to 0.3ms).

Turing had always intended to provide even quicker access to vital data, which is where a special cluster of shorter mercury delay lines came into play. "By good programming, the full advantages of the large store can be enjoyed, with a time access not much greater than is given by the short delay lines," reads an anonymous paper entitled 'A Simple Guide to ACE'.[53] It later noted: "In many cases, the computer will average 15,000 operations per second, each involving extracting two operands,[54] storing the result and extracting the next instruction."

In the meantime, at the General Post Office, a team led by William Chandler and Allen Coombs had carried on developing its version of the ACE after the NPL had decided its manufacturing services were no longer needed. One of Chandler and Coombs's biggest decisions was to use Version VII of Turing's design, and they

---

[53] As above, p80
[54] An operand is a quantity or value on which an operation is performed. So if you were to add two numbers, x and y, x and y are the operands.

came far closer to meeting his storage demands with the MOSAIC (loosely standing for Ministry of Supply Automatic Computer). This included 64 long delay lines and 7000 valves. MOSAIC was a classified project, but it's believed to have been in active service from 1955 until the early 1960s.

The Pilot ACE kept running until 1955, at which point it was replaced by DEUCE. Which made sense, as the DEUCE was essentially a replica of the Pilot ACE but built to commercial standards by the English Electric Company. We will talk more about this in Chapter 11. DEUCE stands for 'Digital Electronic Universal Computing Engine' and it proved a commercial success, in the context of its era, selling over 30 units.

Three years later, in late 1958, came a far closer approximation to what Alan Turing had laid down on paper in 1945. 'The Big ACE' may not have been an imaginative name, but it was apt: this 6000-valve machine "filled a room the size of an auditorium"[55] and ran at 1.5MHz, although surely Turing would have been disappointed to see that it only featured 24 long delay lines. At least the storage drums had been upgraded, now rotating every 5ms rather than 7ms.

Harry Huskey also makes a surprise reappearance in the ACE's history, having designed a commercial computer that was built by American company Bendix in 1954. The size of a large wardrobe, some have called the Bendix G-15 the first personal computer, but at a price of around $50,000 that seems optimistic – the advertising suggests it's "for engineering and scientific computation" and customers included civil engineering firms and universities. It was successful, too, with over 400 computers sold and Bendix even shipping a variety of accessories (such as the PA-3 pen plotter).

But Alan Turing was never to see this or any other successors to the Pilot ACE. He died from cyanide poisoning on 7 June 1954, at the height of the Pilot ACE's commercial success.

While there has been speculation that this was suicide from eating an apple infused with cyanide, based on the presence of half an apple, the apple was never tested for cyanide. What's more, he was experimenting with the compound in a neighbouring room at the time, and toxic fumes from this may have been the real killer.[56]

---

[55] Jack Copeland (ed.), *Alan Turing's Automatic Computing Engine*, p77
[56] For more information, read Chapter 4 of *The Essential Turing*, by Jack Copeland.

The suicide theory was given credence by the fact Turing had undergone a year's worth of hormone therapy, as ordered by the judge in a trial where he was convicted of 'gross indecency' in 1952. Gross indecency being code for homosexuality, then illegal in Britain. But Turing had completed the course more than a year previously, and according to friends he had done so with good humour.

We shall never know the true reasons behind Turing's death.

It's easier to surmise why his elegant architecture, based on optimum coding and minimal hardware, did not survive beyond the computers listed here. As valves were replaced by silicon, hardware's capacity to complete operations hugely increased and the need for streamlined programming lessened. Some might argue that it was dealt a terminal blow by the success of the Williams-Kilburn tube in the Manchester Baby, which made it possible to instantly access memory rather than wait for it to emerge from mercury storage delay lines.

But 'terminal' is too strong a word. In fact, the idea of optimum coding lives on in the power-efficient RISC architecture that our phones run on. Meanwhile, Turing's contribution to the science of computing – most notably his work on artificial intelligence – has never been more relevant than it is today.

BTM HEC 1 computer in the Birmingham Museum Collection Centre
Image: Geni, CC BY-SA 4.0

What happened next

There is no more dramatic event on TV than the calling of a presidential election. The graphics flash, the host looks directly into the camera lens, pauses, then the historic words emerge: "We can now announce that the next president of the United States of America is…"

In November 1952, the end of that sentence finished with the words "Dwight D Eisenhower". And the first person to realise was standing in front of the UNIVAC. For months, a team from Remington Rand – which had bought the Eckert-Mauchly Computer Company in 1950 (see Chapter 9) – had been working with CBS to create a computer program that could transform early voting data into a decisive projection.

"If the parts meshed properly and the program was properly written, we could in all probability announce the winner of the presidential race while our competitors were still floundering in a sea of unsorted data," wrote CBS executive Sig Mickelson[1] in his memoir, glee in his every word, as he recalled the first time the idea was raised.

At the time that CBS and Remington Rand agreed the contract, though, there were many unknowns. How would they create the program? Could UNIVAC work quickly enough? Would the results be accurate?

The answer turned out to be a resounding yes, with the computer correctly calling a landslide victory to Eisenhower with only three million votes in, according to Remington Rand's Arthur Draper on the night. But humans overruled the computer, pretending it had merely given an odds-on prediction in favour of Eisenhower. "We just plain didn't believe it," Draper told the watching audience in an on-air confession later in the broadcast[2].

"As more votes came in, the odds came back, and it was obviously evident that we should have had nerve enough to believe the machine in the first place," he explained. "It was right, we were wrong."

Despite this very human intervention, the UNIVAC's election success was yet another reason for American business executives to excitedly consider the future role computers might play in their companies. The ground was already fertile, with

---

[1] Sig Mickelson, *From Whistle Stop to Sound Bite: Four Decades in Politics and Television* (Praeger, 1989, ISBN 978-0275926328), p138

[2] After 3.4 million votes had been counted, UNIVAC called the result as 438 electoral votes for Eisenhower versus 93 for his rival Adlai Stevenson; the actual result was 442 versus 89. For the long but fascinating story of UNIVAC's role on election night, read Ira Chinoy's PhD dissertation 'Battle of the Brains: Election-Night Forecasting at the Dawn of the Computer Age', University of Maryland, 2010. The stats and Jack Draper's direct quotes are from this paper, which can be downloaded from **drum.lib.umd.edu/items/f5572507-e70d-47c9-ba89-68631e22efc2**

story after story in newspapers about the latest electronic brain, whether it was the UNIVAC, the Princeton IAS computer, or the myriad other magical inventions.

For instance, while many people know about UNIVAC's role in the CBS election night coverage it actually had a rival in the Monrobot over on NBC. This was much smaller, the size of an office desk, and considerably less powerful, but it too tipped Eisenhower as the winner relatively early in the proceedings. The coverage also served as an advert for Monroe's new, small-scale computer, targeted at businesses.

Monroe was unusually early to enter this market, but you can see why it was keen. It had an established network of salespeople to sell its bookkeeping and accounting machines, and a digital computer that could do it all made for an excellent upgrade.

One of its biggest competitors was also eyeing up this new opportunity. "By the early 1950s," wrote IBM historian James Cortada,[3] "while executives and engineers from IBM and the Eckert-Mauchly Computer Corporation (EMCC) were speaking with public officials to drum up business for computers, middle managers and senior executives in the private sector were beginning to sponsor internal studies on the potential uses of computers. My own studies of how more than 30 private-sector industries adopted computers failed to turn up an exception to this pattern of behaviour."

Little wonder that IBM overcame its initial reservations about digital computers and embraced them with a stream of releases that decade. IBM announced the room-sized Model 701, also known as the Defense Calculator, in 1952, with the more business-focused Model 702 to follow three years later. Both used vacuum tubes for speedy calculations and an adapted version of the Williams-Kilburn CRT (as used in the Manchester Baby) to store information.

More Model 700 series computers followed, but it was the IBM 650 that proved the breakthrough. Launched in 1954, the 'Magnetic Drum Data Processing Machine' would be the first computer to sell in the thousands. Although 'sell' is the wrong word: IBM rented them to businesses for around $3500 per month, and is said to have donated around 100 machines to universities in return for them running computer science courses.[4] Its not-so-hidden hope being that the next generation of executives

---

[3] James W Cortada, 'The ENIAC's Influence on Business Computing, 1940s–1950s', in *IEEE Annals of the History of Computing*, Vol 28, Issue 2, 15 May 2006
[4] Donald Knuth quoted in 'The IBM 650', **ibm.com/history/650**

and scientists would lean towards IBM machines, as that is what they had learned on. Apple, Google, and Microsoft all adopt similar tactics today.

IBM needed to be aggressive as it faced plenty of opposition during the 1950s. The UNIVAC was such a PR success that it became almost synonymous with digital electronic computers. By the end of 1954, its new owners had rolled out around 20 systems at $1 million (roughly) apiece, with customers including DuPont, US Steel, Consolidated Edison, and Pacific Mutual Life Insurance.[5] In October that year, it took over the handling of payroll checks for General Electric, marking a historic shift from punched cards to magnetic tape.

By this time Remington Rand had also bought a Minnesota company called Engineering Research Associates (ERA), founded in 1946 by Howard Engstrom and William Norris. Much like Eckert and Mauchly, ERA's battle to make ends meet led to the company being snapped up. Its beginnings were in building code-breaking components for the US Navy, based around early magnetic drum technology that ERA invented, and in 1951 the company developed a computer around it: the Model 1101.[6]

The drum proved a bigger hit than the computer, especially as the technology improved to hold more words, which meant IBM's competitive field of companies was spreading to smaller firms which were using ERA's drum to build computers that cost around $30,000 – equivalent to around nine months of rental for an IBM 650.

There were other rivals too. As featured in the story of the Pilot ACE (Chapter 10), Bendix even produced a "personal computer" called the G-15; personal in the sense that a single scientist might use it, rather than for typing letters. The G-15 was designed by Harry Huskey along the lines of Alan Turing's ACE (most other computers here fall broadly under the von Neumann architecture) and shipped around the world.

None of these companies could sell in the same volume as IBM, however, with figures in the low hundreds at best. This is why the market was eventually known as IBM and the Seven Dwarfs (the unflattering latter label being applied to Burroughs, Control Data, General Electric, Honeywell, NCR, RCA, and Sperry Rand).

---

[5] Thomas Haigh and Paul E Ceruzzi, *A New History of Modern Computing*, Kindle edition, page 24
[6] Converting from binary to decimal, that means Model 13 – building a computer was the 13th 'task' that the US Navy set ERA, and the Model 1101 was its commercial spin-off.

Along with the commercial enterprises, several university-driven computer projects that began in the 1940s continued into the early 1950s, with perhaps the most notable being the Whirlwind Project at MIT. A project that laid the foundation for computer storage for the next two decades.

It stemmed from Jay Forrester, who was then Associate Director of the MIT's Servomechanisms Laboratory. He led a project to create a real-time flight simulator for the US Navy, but the early version was based on analogue technology and didn't work accurately enough. After gatecrashing the Moore School lectures in the summer of 1946, Forrester realised that stored-program digital computers were the best way forward and created Project Whirlwind to test the concept.

Using a simplified design, the team completed the Whirlwind's logical design in 1947[7] and went on to build it using 32 CRTs for storage. The machine slowly evolved into being, from running successful tests in 1949 to fully working in 1951. In terms of specification, it wasn't special for the time other than a 16-bit word length (shorter than the 40 bits typical of its rivals). What made it special was an upgrade to the storage media.

"As various early digital computers were being developed, the characteristics of the available information storage tended to determine the design of the computing machine," said Forrester.[8] In 1947, at the time he was thinking about this, storage came from ERA's magnetic drum, CRT storage as developed by Williams and Kilburn in Manchester, and the still dominant mercury delay line storage. While MIT had created its own CRT storage tubes, they were unreliable – unlike the rest of the machine, which was built for ease of maintenance and long running times.

Forrester's first insight was that the storage media should work in three dimensions rather than two. But what kind of 'flip-flop' device, one that could represent 0 or 1 through the application of current, could they use? Early experiments with glow-discharge elements (such as neon bulbs) proved unreliable, but an unlikely source of inspiration came through an advert for 'Deltamax' in a copy of *Electrical Engineering* magazine. This rectangular loop of magnetic material could be placed into two states,

---

[7] Robert R Everett, 'Whirlwind', found in *A History of Computing in the Twentieth Century* edited by N Metropolis, J Howlett, and Gian-Carolo Rota (Academic Press, 1980, ISBN 978-0124916500), p367

[8] 'Conversation: Jay W. Forrester', interview with Christopher Evans, in *IEEE Annals of the History of Computing*, Vol 6, Issue 3, Jul–Sep 1983, pp297-301

The Whirlwind computer at the Museum of Science, Boston
Image: Daderot, CC BY-SA 3.0

one with high inductance, so current couldn't flow through it, and another where it could. In other words, 0 or 1.

Experiments showed that Deltamax was too fragile, but the idea itself had merit. Switching their focus to ceramic ferrites (a material primarily made from iron oxide that's highly attracted to magnetic fields) proved the pivotal next step. "There was a progression from 1949 over the next two or three years, starting with a single core, testing it, going to 2×2 rectangular arrangements of cores, then gradually to larger and larger planes up to 16×16, and eventually 32×32 arrays while at the same time testing cores to find out how they functioned and to develop higher speeds," said Forrester.[9]

---

[9] Independently, Dr Andrew Booth at Birkbeck College, University of London who will enter our story shortly, and Jan Aleksander Rajchman of the RCA came up with similar ideas.

Ken Olsen, then a graduate student at MIT and later the co-founder and president of Digital Equipment Corporation, suggested they build a smaller version of Whirlwind – the cleverly titled Memory Test computer – to see if magnetic core memory could work in real conditions. It did, and within two years the MIT had upgraded Whirlwind with faster and more reliable memory. The computer carried on its loyal service for the US Navy until 1959.

So far we have stayed in the USA, but the 1950s was also a fascinating period in British computing. We have to begin with LEO I, which is undoubtedly the first computer to be bought by a teashop.

If you insist on accuracy, a chain of teashops, one that stretched the length of the country at the time, owned by J Lyons & Co. With over 200 stores to serve, and an emphasis on affordable prices and speedy service (the so-called 'Nippy' waitresses were famous for this), the company was constantly striving for efficiency. So much so that it had set up its own Systems Research department to investigate new ideas and technology, with a Cambridge mathematics graduate at its helm: John Simmons.

One of Simmons's wilder pre-war ideas was for an advanced automatic calculator. This, he explained, would store data in its magnetic records that could then be called upon by the machine they already used for accounting. He showed his sketched-out plans to the company's chief engineer, Jack Edwards, but that was as far as the idea went before war intervened.

It was at least partly Simmons's concept that drove Thomas Raymond Thompson and Oliver Standingford, who both worked with Simmons in the Research Department, to travel to America in early 1947 and report back on the latest electronic developments happening there.

In her excellent book, *A Computer Called LEO*, Georgina Ferry describes the pair's first meeting with Herman Goldstine, now working with John von Neumann on the Princeton IAS computer (see Chapter 9). The Englishmen explained how they hoped to use a computer as part of their office systems, which fundamentally involved processing data. Goldstine hadn't considered this idea before but proved quick on the uptake. "Sketching furiously on a yellow pad, [Goldstine] launched into a description of possible approaches to the problem given the technology that had been developed so far," Ferry describes.[10]

---

[10] Georgina Ferry, *A Computer Called LEO* (Fourth Estate, 2003, ISBN 978-1841151854), p43

Goldstine went on to explain the benefits of stored-program computing to his guests and gave them a list of people to visit (including Presper Eckert). "Then, enjoying the astonishment of his listeners," writes Ferry, "he dropped his bombshell. 'And, of course, there's Professor Douglas Hartree in Cambridge, England.'" They had travelled across the Atlantic ocean, when they could have driven to the university in little over an hour.

Goldstine dashed off a letter of introduction for the duo, and when Edwards and Standingford eventually met with Maurice Wilkes, the true power behind Cambridge's EDSAC project rather than Hartree, the meeting could not have gone better. It helped that the Lyons men, like Simmons, were both top-class mathematics graduates from Cambridge University.

As mentioned in Chapter 8 on the EDSAC, Lyons would eventually back the project with a cheque for £3000 (roughly £90,000 or $120,000 in today's money) and the free gift of an engineer until the Cambridge computer was finished. In return, the engineer, would learn all there was to know about the machine and Lyons would be in the ideal position to build their own copy of the computer. A young man named Ernest Lenaerts was more than happy to fulfil this position.

Wilkes would also help in the hiring of Lyons's chief engineer, by hinting to a gifted physics PhD student, John Pinkerton, that he might just want to apply for a vacancy cryptically placed in the science journal *Nature* for "a graduate electronics engineer aged 25-35".[11] Pinkerton proved the perfect fit, starting in January 1949, and by May that year their computer had a name: the Lyons Electronic Office. Simmons, who clearly had a pedantic side, pointed out that the actual computing was done by the arithmetic unit, so the whole machine could not be called a 'computer' itself.

LEO also had a home, on the second floor of an administrative block that was part of Lyons's sprawling HQ, and collection of bakeries, in Hammersmith, London. And so, in a very British way, with the smell of freshly baked bread floating through the windows, one of the world's first commercial computers came to be built. By November 1951 it was put to work calculating the value of each week's bakery output, and in January 1952 it ran a program "repeatedly for 59 hours, during which LEO broke down 14 times, needing a total of three and a half hours for repairs,"

---

[11] Georgina Ferry, *A Computer Called LEO*, p88

wrote Ferry.[12] That gave it an operational efficiency of 87%, which was exceptional for the time.[13]

Most companies would crow about such successes, but Lyons wasn't in this for the publicity. Still, the world of building computers/arithmetic units was a small one in 1950s Britain, and the Ministry of Supply requested time on the computer to calculate ballistics trajectories – exactly what the ENIAC was designed for – in return for £300. Thompson, one of the original duo who went to America and who was now in charge of LEO day to day, agreed. But only, he said, because it would "provide us with experience in operating the machine on a different kind of work".[14]

It would take until October 1954 for newspaper readers to discover that "Electronics join the staff", to quote the report's headline. "L.E.O. (Lyons Electronic Office) now calculates in one afternoon the pay-packets of thousands of employees – work which previously kept an army of clerks busy for the whole week," the article explained. "So efficient indeed is L.E.O. that Lyons cannot keep it fully occupied: it is hired out to other firms for spare-time work and has also been called into serving the War Office and the Air Ministry."[15]

A year later, tucked away in *The Daily Telegraph*, came the curious news that, at the previous day's annual general meeting for Lyons, "Mr M Gluckstein, the chairman, told shareholders that the company are 'in the market to sell calculating machines.' He added: 'We have a computer of our own which we call Leo, and we are having another brought into use. We have had certain inquiries but nothing much has happened so far.'"[16]

Readers of Chapter 7, on the development of the Manchester Baby, will know that Ferranti was installing its Mark 1* – based upon the University of Manchester's Mark 1 design – in countries across Europe by the mid-1950s. In a similar vein, in

---

[12] As above, p109
[13] Mary Coombs, one of the early programmers on the LEO, explained the difficulty of diagnosing errors in a 2010 interview. "So we spent many evenings sitting with the engineers, because the only way to find out which valve wasn't working was to try and find out where in your program it went wrong, so you could pinpoint the rack that held the instruction or the data where it was going wrong. And this could take hours." Especially one problem where they just couldn't determine the reason for an intermittent fault. "We eventually discovered that the management lift, which went up to the fifth floor where the boardroom was, was interfering." The interview was by Thomas Lean forms part of 'An Oral History of British Science', shelfmark C1379/16/01-09
[14] Georgina Ferry, *A Computer Called LEO*, p105
[15] T Davenport, 'Electronics join the staff', in *Evening News and Star* (late final edition), 14 October 1954, p8
[16] 'Lyons computers for sale', in *The Daily Telegraph*, 6 July 1955, p2

1955 industrial manufacturer English Electric started shipping its own version of the Pilot ACE, the DEUCE, with roughly 30 sold overall.

These were big installations, like the LEO, but there are two other small computer projects worthy of note. The first started life at Birkbeck College, part of the University of London, in the late 1940s, where Dr Andrew Booth devised a small computer – that he called the APE(X)C.[17] This used around 500 vacuum tubes to solve scientific problems and featured a magnetic drum for storage.

At this point, the British Tabulating Machine Company was desperately in need of a computer. For years, it had benefited from an agreement with IBM where BTM sold replicas of IBM's machines in return for a 25% royalty paid to the US company. This contract ended in 1949. So BTM hired John Womersley, who played an important role in the creation of the Pilot ACE, to build its new computers. Womersley heard of Booth's work and despatched Raymond Bird "to a rotten barn in a village called Fenny Compton where Doc Booth was developing the prototype of his APE(X)C machine", said Bird.[18]

Over three months, Bird and two colleagues painstakingly copied the plans for the machine. "The chronology of all this is that I joined BTM on 1 January 1951, shivered in Booth's barn copying his plans during March of that year, and had built the prototype and got it to work by the end of 1951," Bird recalled. The end result, complete with office-friendly, IBM-style tabulators, was called the HEC 1. The production version, HEC 2, featured a larger drum and a more professional design.

More versions would follow, along with new names: the 1201 with a 1024-word capacity and the 1202 that stored 4096 words. "We sold, or at least we delivered, 125 1201s and 1202s," said Bird. "That was more than all the other contemporary British computers put together, so it was a significant achievement." Much of its success was to do with the price (at £30,000 it was a third of the cost of the LEO, for example) and that it included the punch-card equipment that so many of BTM's existing customers relied upon.

---

[17] This, rather clumsily, stands for 'All Purpose Electronic (X-initial of sponsoring agency) Computer', according to Booth's account in *Mathematical Tables and Other Aids to Computation*, Vol 8, No 46, April 1954

[18] Raymond Bird, 'BTM's First Steps Into Computing', in *Resurrection: The Bulletin of the Computer Conservation Society*, No 22, Summer 1999, ISSN 0958-7403, transcript based on talk given to the Punched Card Reunion in Stevenage, Hertfordshire, England, in 1998 (according to The Register, rpimag.co/raymondbird)

To complete the British picture, in 1953 a company called Elliott Brothers, based in the town of Borehamwood a few miles north-west of London, surprised everyone by releasing its Elliott 401. It was modest by Mark 1* standards, with only 4 kilobytes of memory and a lowly clock rate of 333kHz, but it also cost a third of the price and was designed for mass manufacturing.

Barring improvements to existing designs – for example, the more compact and slightly faster LEO II came out in 1957 – this provides an overview of the computer market in the UK and the US by the end of the 1950s. Both markets are notable for their fragmented nature, a mixture of what might be considered startups with established equipment manufacturers.

Inevitably, what happened next was consolidation. Crises leading to mergers as everyone attempted to fend off the might of IBM.

But it's tempting to consider what might have happened if Ferranti had been stronger. Or if Remington Rand, which merged with the Sperry Corporation in 1955 to form Sperry Rand, had more visionary sales teams. "It was unbelievable, unbelievable, when they took over," said Betty Holberton of the Remington Rand sales team after they bought out Eckert and Mauchly.[19] "It was just like having a sales group come in and take over something for which they didn't have any idea what they were going to do with it or what direction they really wanted to go."

It's a thought backed up by Jean Bartik's experience. "I worked for them in 1950/51 in the Washington sales office," she said. "I was the only one there that knew anything about the UNIVAC. Well, their salesman used me to make sales of typewriters, would you believe. They got their foot in the door with me talking about UNIVAC, this exciting new machine, and then they quietly on the side sold a few calculators and bookkeeping machines or typewriters. Because their salesmen didn't know a thing about UNIVAC."

Sperry Rand also struggled to take advantage of John Mauchly's undoubted talents. He wasn't a natural fit for a big corporation and left the company in the late 1950s after it asked him to move to Washington as part of its sales team. Presper Eckert stayed, but his influence waned as time moved on.

---

[19] Oral history interview with Jean J Bartik and Frances E (Betty) Snyder Holberton by Henry S Tropp, 27 April 1973, Smithsonian National Museum of American History, **rpimag.co/bartikholbertoninterview**. The quotes have been edited in places to avoid repetition.

Michael Godfrey, who was director of research at Sperry Rand from 1976 to 1983, wonders if things might have been different if Mauchly had remained with the company. "I talked with the people who knew him well," he said.[20] "They all liked Mauchly, and from what they told me he was the one who actually tried to get things done."

And while people may think that IBM was indomitable, it didn't always have the best computers, Godfrey believes. "IBM really blundered with the System/360," said Godfrey. "Once a programme was loaded into a fixed memory address, it could be swapped out, but it had to be put back where it came from. There was no relocation in the machine. So essentially you had to run one job at a time."

For giant engineering firms such as Boeing, this was terrible news. So when the time came for them to replace their existing IBM machines, they looked around for alternatives such as the Ferranti Atlas – which could be considered a supercomputer of the time.

"That's when a friend of mine from Boeing, who looked after engineering computation, called Ferranti," recalled Godfrey. "He said, I have 20 friends who will buy your machines and install them in development labs around the US. All we ask is you have some support staff available to support the machines in the US. And Basil de Ferranti said, 'you can always call us in Manchester. Goodbye.' And that is one of the singular leaps backward of all time."

Away from hardware, we also saw the growth of programming as a discipline. In 1951, Addison-Wesley published *The Preparation of Programs for an Electronic Digital Computer* by Maurice Wilkes, David Wheeler, and Stanley Gill, based on their experience of programming the EDSAC. It would popularise, if that's the word, the concept of debugging, introduce APIs, and talk about reusing libraries of code. The book even earned its own nickname of WWG and would influence programmers in the UK, the US, and worldwide.

But the big push, in terms of making computers widely usable, took the creation of high-level languages: the arrival of FORTRAN in 1957 for scientific computing and COBOL in 1959 for business applications are standout releases.

Meanwhile, the outlines of a worldwide computer industry was emerging. Outside of Britain and the USA, Australia was arguably the quickest off the mark

---

[20] Interview with author

The control panel of the BESK computer at the Tekniska Museet in Stockholm
Image: Liftarn, CC BY-SA 3.0

with its CSIRAC (based on the EDVAC design) running its first test program in 1949 – although it wasn't publicly demonstrated until 1951. It kept working until 1964, and unlike most other electronic computers of the era it wasn't broken up: it's on permanent display at the Scienceworks museum, Melbourne, as part of its Think Ahead exhibition.

In mainland Europe, Switzerland's adoption of Konrad Zuse's Z4 in 1950 gave it an instant head start. Sweden proved similarly keen, producing its relay-based BARK in 1950 and following it up in 1953 with the fully electronic BESK.

Despite the absence of Zuse, who supervised the Z4 in Switzerland, West Germany also continued its pioneering work. Heinz Billing created a stored-program computer using vacuum tubes and drum storage called the G2 in 1954, following it up with the larger G3 in 1960 (the G1 was a test machine rather than a full computer). By the end of the decade, Germany would be home to two more large-scale computers: one

The Ural-1 computer at the Polytechnic Museum, Moscow
Image: Panther, CC BY-SA 3.0

based on the Princeton IAS design, the PERM, another more influenced by Howard Aiken called DERA.

The USSR was even busier, with the MESM – standing for Малая Электронно-Счетная Машина, or Small Electronic Calculating Machine – completed in 1950. The BESM-1, completed in 1952, was much faster (the B stands for Большая, or 'bigger'), and would be succeeded by five further models. The government also set up the Ministry of Radiotechnical Industry to turbocharge research, with three more computers – the Ural-1, Ural-2, and Minsk-1 – all appearing before the end of the decade.

Then there is the small – tiny, in fact – matter of the transistor. As we mention in the story of George Stibitz's Complex Number Calculator (Chapter 3), Bell Labs developed this in the late 1940s. The transistor spelled the end for big, expensive, power-hungry vacuum tubes. Within a decade, the integrated circuit had arrived, and we were on the road to the computers that we would recognise today.

Gordon Moore introduced a concept that became known as Moore's Law in 1965. "I was given the chore of predicting what would happen in silicon components in the next ten years," he recalled in 2012[21] as he remembered writing the legendary

---

[21] Gordon Moore, 'Excerpts from A Conversation with Gordon Moore: Moore's Law', Intel, 2012, **rpimag.co/mooreintel**

244 The Computers That Made The World

article for *Electronic Magazine*. To help him gauge the rate of future growth, he counted back to see how quickly developments had happened in his own work. "I said gee, in fact from the days of the original planar transistor, which was 1959, we had about doubled every year the amount of components we could put on a chip," he said. From this he "blindly extrapolated for about ten years and said, okay, in 1975 we'll have about 60,000 components on a chip".

That assumes a doubling every year, a prediction that held for a decade. At that point he revised his prediction to a doubling every two years, although actual increases outpaced this until the early 2020s. This is shockingly quick: if cars had improved at the same rate, they would travel at roughly 50 trillion miles per hour.

Since Moore made that famous prediction, we have seen the worldwide economy rewritten by the World Wide Web. The explosion of social media. The arrival of smartphones that are now countless times faster than the ENIAC. The rise of artificial intelligence in almost precisely the way that Alan Turing foretold.

And to think, it all started from people wanting to add things together.

The Computers That Made the World
TIMELINE

1951

Maurice Wilkes, David Wheeler, and Stanley Gill write the first book on programming

Ferranti Mark 1
Production version of the Manchester Mark I delivered to Manchester University

Princeton IAS
John von Neumann's computer, based on his famed architecture, goes online

MIT Whirlwind 1
Huge military computer completed with revolutionary magnetic-core memory

UNIVAC 1
Eckert and Mauchly's computer delivered to US Census Bureau

LEO 1
Britain's J Lyons & Co embraces EDVAC-style computer to help streamline its restaurant, food, and hotels business

1953

IBM 701
The 'Defense Calculator' appears, marking the beginning of IBM's mainframe era

1954

DEUCE
Commercialised version of the Pilot ACE goes on sale

IBM 650
First mass-produced computer, selling almost 2000 systems during lifetime

1955

October: the ENIAC is finally shut down

1956

Los Alamos chess, a variant on a 6×6 board with no bishops, is developed and beats human player

1957

IBM releases Fortran, one of the earliest high-level programming languages

COMMERCIALISATION

PROGRAM

1958

- **ALGOL language invented; COBOL would follow a year later**

- **RCA 501**
  Early transistorised computer goes on sale

1960

- **Invention of the integrated circuit**

- **PDP-1**
  A family is born as DEC's PDP-1 goes on sale

1963

- Tony Brooker and Derrick Morris produce Atlas Autocode

1964

- **BASIC language invented**

- **CDC 6600**
  Using 400,000 transistors, this is 3× faster than any rival computer

1965

- Gordon Moore writes article speculating that components on integrated circuits would double each year for a decade – Moore's Law

- **IBM System/360**
  First System/360 shipped, marking another big shift in computing history

SUPERCOMPUTERS

Index

A
ABC 2-19, 48, 89, 110, 113, 118, 126, 127
Aberdeen Proving Ground 54, 120, 191
Aiken, Howard 31, 36, 86-107, 155, 189
Analytical Engine x-xvii, 24, 90, 104, 218
APE(X)C 240
Atanasoff, John 2-19, 48, 89, 92, 113, 126, 127
AT&T 38, 40, 41, 42, 43, 44, 50
Auerbach, Isaac 166, 182, 183, 184, 185, 197
Autocode 148
AVIDAC 196

B
Babbage, Charles x-xvii, 24, 83, 90-91, 143, 208
BARK 243
Bartik, Jean 111, 121, 122, 124, 125, 185, 197, 241
Bell Labs 38-57, 63, 131, 132, 168, 178, 180, 181, 189, 216, 218, 244
Bennett, John 148, 155, 159, 160
Berry, Clifford 2-19, 127
BESK 243
BESM-1 244
Bigelow, Julian 192-196
BINAC 112, 158, 166, 175, 183-187, 190, 191, 198, 199
Bletchley Park 58, 60, 63-81, 83, 84, 133, 135, 164, 204, 209, 212, 213, 215-217
Bloch, Richard 102
Bomba 209-211
Bombe 211, 212, 213, 215, 216
British Tabulating Machine Company 211-212
BRL (Ballistic Research Laboratory) 110, 111, 113, 114, 120, 125, 168, 169, 171, 190, 191
Broadhurst, Sid 60, 74
Bryce, James 94-96, 101
BTM (British Tabulating Machine company) 240
B-tube 144
Burks, Arthur 7, 13, 115, 116, 120, 124, 125, 172, 174, 178, 180, 187, 188, 192, 193
Bush, Vannevar 114, 153, 156, 176

C

Carty, John 40, 41
Chambers, Carl 119, 187, 188, 190
Chandler, Bill (William) 60, 74, 76, 77, 79, 219, 227
Chapline, Joseph 111, 112
COBOL 242
Colossus 58-85, 100, 104, 133, 135, 168, 216-219
Complex Number Calculator 38,-57, 168, 244
Coombs, Allen 79, 82, 219, 220, 227
CRT 131-133, 137, 140, 142, 178, 194, 195, 233, 235
CSIRAC 243

D

Deltamax 235-236
DERA 244
DEUCE 228, 240
Difference Engine x-xvii, 90, 143
Dollis Hill 63, 74, 75, 79, 83, 216, 219
DVL (Deutsche Versuchsanstalt für Luftfahrt) 29, 31, 33

E

Eckert, Presper 7, 15, 106, 113-120, 123-127, 131, 152, 155, 156, 170-190, 196-199, 232-233, 234, 238, 241
EDSAC 36, 131, 135, 148, 150-165, 175, 189, 209, 217, 219, 227, 238, 242
EDVAC 36, 82, 105, 123, 127, 152, 154-158, 167, 168, 170, 172-174, 176, 181, 187, 188, 190, 191, 196, 201, 207, 217, 219
EMCC (Eckert-Mauchly Computer Corporation) 198, 199, 233
ENIAC 6, 7, 12, 14, 15, 19, 22, 32, 46, 50, 54, 81, 82, 84, 92, 97, 100, 103, 108-127, 131, 152, 154, 155, 168, 169-174, 176, 178, 179, 181, 182, 184, 187-189, 191, 197, 198, 218, 221, 233, 239, 245
Enigma 64-66, 204, 208-216
Entscheidungsproblem 204-206
ERMETH 36
ETH 35-37

F

Ferranti Mark 1 143, 145-148
Ferranti Mark 1* 147, 148, 239, 241
First Draft Report (on EDSAC) 172, 179, 188, 192
Flowers, Tommy 60-64, 71-76, 79, 83, 84, 135, 217, 219, 220

Forrester, Jay 235
FORTRAN 242

G

General Electric 7, 43, 51, 189, 234
Gill, Stan 162
Goldstine, Herman 110, 111, 112, 113, 114, 115, 116, 117, 118, 120, 123, 124, 125, 137, 156, 168, 169, 170, 171, 172, 174, 175, 176, 178, 179, 187, 189, 192, 237, 238
Gold, Tommy 156, 158
Good, Jack 74, 135, 139
GPO (General Post Office) 60, 62, 71, 79, 135, 168, 221, 227

H

Hamilton, Frank 94-97, 100
Hamming, Richard 55
Hartree, Douglas 134, 154, 155, 172, 175, 187, 189, 218, 219, 221, 238
Harvard Mark I 6, 26, 31, 35, 52, 86-107, 114, 155, 168
HEC 1 240
Holberton, Betty 121, 122, 124, 125, 184, 185, 197, 241
Hopper, Grace 101, 102, 106, 220
Huskey, Harry 187, 221-223, 228, 234

I

IBM 4, 5, 6, 7, 31, 90, 92-103, 105, 114, 117, 118, 169, 180, 181, 189, 196-199
IBM 650 233, 234
IBM 701 196, 233
IBM System/360 242
ILLIAC 196

J

Jewett, Frank 39, 40, 41, 43, 44
JOHNNIAC 196

K

Kelly, Mervin 43, 56, 152, 153, 189
Kilburn, Tom 130, 131, 135-141, 143-145, 147-149, 195, 229
Knox, Dilly 65, 209, 211

L

Lennard-Jones, John 153
LEO I 237

Los Alamos 82, 103, 104, 120, 123, 124, 168, 169, 193, 195, 196
Lovelace, Ada x-xvii, 208
Lukoff, Herman 184, 187-190, 196, 197, 199, 200
Lyons (J Lyons & Co) 159, 161, 237, 238, 239

M

Magnetic drum 234, 235, 240
Manchester Baby 80, 82, 128-149, 152, 223, 229, 233, 239
MANIAC 196
Mauchly, John 7, 12, 14, 15, 22, 50, 82, 92, 106, 110-114, 116, 118, 120-127, 152, 155, 160, 170, 172-174, 176, 178-182, 184, 185, 187-190, 196-199, 232-234, 241, 242
Mercury delay line 156, 158, 171, 187, 222, 227, 235
MESM 244
Michie, Donald 74, 79
Millikan, Robert 41-43
MIT 36, 40, 50, 56, 88, 93, 94, 102, 130, 154, 156, 164, 169, 172, 176, 180, 189
Model K Adder 45, 46
Moore, Gordon 235, 244, 245
Moore's Law 244, 245
Moore School 111, 112, 117, 119, 120, 126, 127, 131, 135, 152, 154, 157, 168, 170, 172, 174-176, 179, 180, 181, 187-192, 196
Morrell, Francis 71
MOSAIC 79, 228

N

Napier's bones 90
Newman, Max 60, 71, 73, 77, 80, 82, 133-135, 138, 144, 145, 147, 176, 204, 205, 207, 211, 217, 224
Northrop 182-184, 186
NPL (National Physical Laboratory) 132-134, 138, 217-224, 226, 227

O

ORACLE 196
ORDVAC 191, 192, 196

P

PERM 244
Pilot ACE 46, 79, 131-133, 135, 152, 176, 187, 202-229, 234, 240
Princeton IAS 138, 147, 167, 171, 176, 191-196, 201, 233, 237, 244

R

Rajchman, Jan 178, 189
RCA 92, 115, 137, 178-180, 194
Remington Rand 198, 199, 232, 234, 241
Renwick, Bill 150, 155, 158, 160, 161
Rockefeller Differential Analyzer 114
Royal Society 90, 130, 133-135, 138, 141, 145-147, 152, 156, 209

S

Schreyer, Helmut 27
Selectron 137, 147, 178, 192, 194, 195
Shannon, Claude 189, 216
Sharpless, Thomas Kite 115, 187, 190
Shockley, William 56, 57
Simmons, John 237, 238
Smithsonian National Musuem 14, 89, 90, 95, 97, 99, 102, 118, 127, 185, 196, 197
Sperry Rand 6, 7, 181, 234, 241, 242
Stecker 210, 212, 214
Stibitz, George 44-54, 57, 79, 169, 176, 181, 188, 216
Stiefel, Eduard 35-37

T

Tiltman, John 65, 66, 68
Tootill, Geoff 137, 139, 140, 143-145
Travis, Irven 73, 77, 174, 180, 187, 211
TRE (Telecommunications Research Establishment) 130, 132, 133, 136, 137, 139, 140, 146
Turing, Alan 22, 37, 64, 69, 70, 79, 82, 130, 132, 135, 138, 142-148, 152, 176, 203, 204-209, 211-229, 234, 235
Turingery 69, 70
Turing machine 206, 217
Tutte, Bill 60, 68-71, 79, 83

U

UNIVAC 106, 146, 167, 176, 179, 181-185, 187, 189, 194, 197-201, 232-234, 241
US Army 53, 56, 110, 112, 118, 121, 122, 124, 152, 157, 168, 170, 173, 176, 179, 196
US Census Bureau 181-183, 187, 189
US Navy 14, 96, 100, 105, 177, 180
US Weather Bureau 179
Uttley, Albert 132, 139

V

Vacuum tubes 10, 14, 16, 27, 72, 83, 114-116, 127, 158, 191, 192
Vail, Theodore 40, 41
von Neumann, John 36, 37, 50, 102, 105, 110, 114, 116, 117, 123, 125, 130, 135, 137, 139, 152, 168-179, 187-189, 191-193, 196, 207, 208, 217, 219

W

Welchman, Gordon 212
Wheeler, David 160-164
Whirlwind 235, 237
Wiener, Norbert 50, 176
Wilkes, Maurice 135, 138, 150, 152-164, 175, 189, 209, 217, 219, 220, 238, 242
Wilkinson, James 220-225
Williams, Frederic Calland 46-48, 50, 51, 71, 82, 129, 130-133, 135-141, 143, 144, 147, 177, 189-191, 195, 223, 229
Williams-Kilburn tube 131, 137, 140, 143, 147, 195, 229
Womersley, John 132, 217-219, 221, 240
Woodger, Mike 220-224
World's first 'stored program' 141
World War II 14, 102, 104, 122, 130, 153, 156
WWG 162, 242
Wynn-Williams, Eryl 71

Z

Zuse, Konrad 20-37, 48, 82, 92, 217
Zuse Z1 24-27, 29-31
Zuse Z3 20-37, 48
Zuse Z4 25, 33-37, 243